# Moving with Math®
EXTENSIONS

 Math Teachers Press, Inc.

Principal Author: Caryl Kelly Pierson
Artist: Ingrid M. Holley
Cover Artist: Bob Klein
Layout and Design: Carolyn Johnson, Bryan Smith, Jim Crinklaw
Readers and Contributors: Linda Farnham, Susan J.P. Gruidl, Lynn Schuster, Michele Dalheimer
Typist: Mary Johnson
Production Editors: Molly E. Hennessey, Ann Bear
Editor: David W. Solberg

© 2003 Math Teachers Press, Inc.
4850 Park Glen Road
Minneapolis, Minnesota 55416

LIMITED REPRODUCTION RIGHTS

The material contained herein is protected from reproduction by Federal and State Copyright laws. All forms of copying and redistribution by any means, including but not limited to photocopies, electronic mail, scanning and posting on the internet or intranet is strictly prohibited and unlawful. The only exception to this strict prohibition of copying and distribution is that the publisher grants to the teacher who utilizes this material a reproduction license for personal use of the reproducible student pages. Reproduction for a school or the district is strictly prohibited. Math Teachers Press retains all right, title and interest in this material and copyrighted material and does not make any assignment of the ownership of any of these materials or documents.

Printed in the United States of America.

ISBN: 1-59167-011-X

Table of Contents

Foreword i
Correlation to Objectives v
Supply List vii
20-Day Lesson Pacing Plan ix
Blank Pacing Plan xiii

Assessment

Student Progress Report 1
Class Record Sheet 2
Pre-Test 5
Post-Test 11
Final Journal Prompt 17
Answer Keys 18
Glossary 25
Journal Prompt Instructions and Scoring Guide 29
Blank Math Journals 31

Teacher Guide

Numeration and Whole Number Operations 1
Problem Solving 14
Fractions 20
Decimals 36
Geometry 50
Measurement 55

Skill Builders Blackline Masters

Masters 1–15
Skill Builders .... 1-1 to 50-2

Moving with Math® Extensions

"We remember 10% of what we hear, 30% of what we see and 90% of what we do."
—Jean Piaget

The *Moving with Math® Extensions* program develops understanding of mathematical concepts through the use of manipulatives, language development and problem solving. This Teacher Manual, when used in conjunction with the student book, provides a condensed program for the grade level. The program is designed to be used as an extension in a variety of settings–extended year summer school, extended day after school or as a supplemental resource. *Moving with Math®* provides all the activities necessary for students to develop fundamental skills and build a knowledge bank.

Theoretical Basis: *Three Stages of Learning*

Moving with Math® utilizes an activity-based approach that mirrors children's learning style. Conceptual learning begins at the concrete level with real-world experience and manipulative activities. Students gradually move from the concrete with manipulatives to a representational mode involving pictures and, then, to a greater level of abstraction where mathematical thinking is represented with language and handwritten signs and symbols. The three stages of learning are illustrated below:

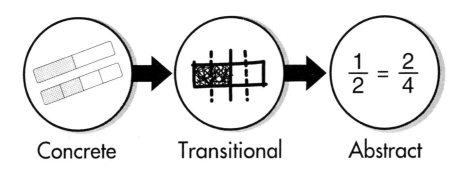

Concrete
The concrete stage consists of actual contact with objects. Students look, touch, and move objects to discover mathematical concepts.

Transitional
In the transitional stage, students bridge the gap between the concrete and abstract with strategies such as drawing pictures and talking about math.

Abstract
Numerals and signs are abstract. Students develop abstract understanding after many experiences in the concrete and transitional stages.

Features of the Program

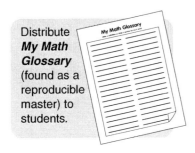

Ask these questions when using manipulatives:
- How are these alike?
- How are these different?
- Is there a pattern?
- Is there a general rule for what we are doing?
- Tell me what to do next.

Using Manipulatives

Learning objectives are introduced at the concrete level with manipulatives. As children explore with manipulatives, they naturally discuss what they are doing and begin to understand the underlying concepts of essential math skills. Ask questions to encourage discussion and assist in this discovery.

Communicating Mathematical Concepts

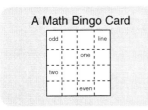

Distribute *My Math Glossary* (found as a reproducible master) to students.

Teaching Math Vocabulary – Before beginning a math lesson, print each vocabulary word on a Vocabulary Card (Master 15). Display the card and read it aloud with the class three times. Mount the cards on a bulletin board or other large area to make a word wall. When the lesson is completed, ask students to write or draw a picture of what the word means in their Math Glossary. (A complete definition of terms is provided in the Assessment Section of this book.)

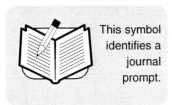

A Math Bingo Card

Math Bingo – This game may be played after at least 16 vocabulary words are on the word wall. Have each student fold a sheet of square paper into 16 parts and write a math word in each part. Write one vocabulary word on each index card. Play Bingo using the index cards to call out words. Students use counter chips to cover words on the 16-part bingo cards.

Journal Prompts – Students need to communicate math concepts both verbally and in writing. Journal Prompts suggest activities for assessment by having students write in a journal or talk about what they have learned. The assignments are specific and should be doable after certain points of instruction have been completed.

This symbol identifies a journal prompt.

Problem Solving

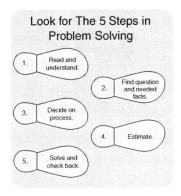

Look for The 5 Steps in Problem Solving
1. Read and understand.
2. Find question and needed facts.
3. Decide on process.
4. Estimate.
5. Solve and check back.

Real-world problems are used as a vehicle for introducing math concepts. Students are encouraged to write their own word problems. Awareness of the following strategies assist children's problem-solving abilities:

- Use a five-step plan
- Use a model
- Guess and check
- Act out the problem
- Make a table
- Write a number sentence
- Draw a picture
- Find the pattern
- Use logic

Playing Games

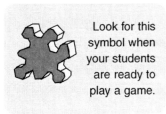

Look for this symbol when your students are ready to play a game.

Research shows that games give students a chance to strengthen math skills and problem-solving abilities. As they develop a winning strategy for a game, they will go through steps similar to those used in problem solving. Always demonstrate how a game is played rather than merely explaining the rules.

Connecting School and Family

Parent Handbooks – Research shows that parent involvement has a positive impact on children's academic achievement. The Family Math Connections parent handbook, based on the principles underlying *No Child Left Behind,* is an important tool for connecting with students' families. The handbook offers math activities for parents and students to enjoy together, describes the Extensions program, and outlines students' strengths and weaknesses.

Connection to Home

Differentiated Instruction

Brain research shows that students who are actively engaged in the classroom learn more and retain the information longer. Active engagement is built into *Moving with Math's* varied and structured hands-on manipulations, games, written activities, journal prompts, literature connections, and teacher-led and student-generated discussions.

The diverse approaches of the *Extensions* program tap into multiple intelligences and learning styles. Different students will find different access points via which they can initially grasp math concepts. And, as lessons move between the concrete, the representational, and the abstract, all students are able to expand and integrate the ways through which they can understand and express essential ideas in math.

ELL Students – *Moving With Math's* use of manipulatives and real-world examples makes lessons accessible to English language learners while strengthening all students' language skills. This is especially true at the primary levels, in which all students are learning basic vocabulary.

Special Needs Students – The program's range of learning activities allows for modifications to meet students' specific learning needs. The varied modes of learning allow for varied types of assessment, so that all students can demonstrate their proficiency.

Individualized Instructional Plans – Teachers can take advantage of the objective-correlated student activity pages, pre-test, and review items to develop individualized instructional plans that allow students to identify their weaknesses and focus their learning.

Lessons tap into diverse learning styles and strengths. Varied modes of assessment allow all students to show what they know.

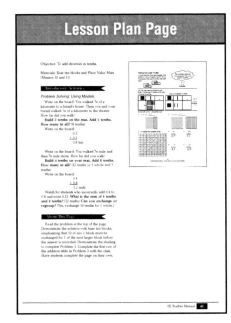

What is inside this manual?

The Teacher Manual contains everything teachers need and provides flexibility for any type of extended school. Page vii provides a list of materials needed for the activities. Pages v and vi contain a diagnostic prescriptive list of objectives correlated to student pages. Pages viii to xiii contain 20-day lesson pacing plans and a blank pacing plan. The rest of the manual is divided into 3 parts:

1. Assessment – Student Progress Report, Class Record Chart, reproducible Pre-Test and Post-Test, Answer Keys, Journal Prompt Instructions and Scoring Guide and Blank Math Journals

2. Teacher Guide – These pages provide lesson plans with lightly scripted, manipulative-based lessons, corresponding to the pages in the student book. After students complete the hands-on activity, they complete the corresponding page. These pages link the manipulative activities to the pictures and to the guided practice at the abstract level. Follow Up Activities include games, suggestions for extra practice, and journal prompts.

3. *Skill Builders* **Blackline Reproducibles**

  A. Masters – Masters such as graph paper, fact tests or game templates may be reproduced for student activities.

  B. *Skill Builders* **–** These blackline masters provide reteaching pages. Each page corresponds to a numbered objective listed on pages v and vi.

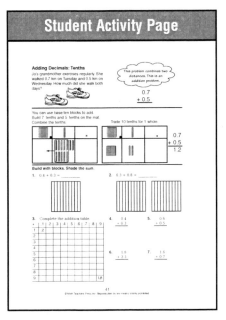

What is inside the student books?

The student books provide in-depth development of the most-needed skills for the grade level. The first 64 activity pages contain problems that follow the hands-on activities for that lesson.

The Daily Reviews found in back of the student book build long-term retention of all grade-level objectives. The inside back cover has a record sheet that correlates review problems to the objective they represent. After the teacher reads the answers, students fill in their record sheets. The students and the teacher can easily identify the missed objective numbers. Then, the appropriate *Skill Builders* page may be assigned.

Correlation to Objectives

Use this table to match pre- or post-test problems to objectives and pages in the teachers manual and student book.

		Numeration	Student Book	Skill Builders
	C-1	Identify the place value in a 7- to 12-digit number.	1	1-1
	C-2	Read, write and compare 9- to 12- digit numbers.	2	2-1
	C-3	Round a 4- or 5-digit number to the nearest thousand.	3	3-1, 3-2
	C-4	Identify prime numbers and the factors of composite numbers up to 100.		4-1
	C-5	Use the commutative, associative or the distributive property.	4	5-1, 5-2
		Whole Number Operations		
	C-6	Add numbers up to 6-digits.	5	6-1, 45-5, 47-1
	C-7	Subtract numbers up to 6-digits.	6	7-1, 45-2, 45-5
	C-8	Multiply a 3-digit number by a 2-digit number. Multiply by multiples of 10.	9, 10	8-1, 8-2, 50-1, 50-2
	C-9	Divide a 6-digit by a 1-digit number.	11	9-1
Test Item/Objective Number	C-10	Divide a 4-digit by a 2-digit number.	12, 13	10-1 to 10-5
		Fraction Concepts		
	C-11	Write fractions from shaded regions, number lines and printed words.	20, 21	11-1, 11-2
	C-12	Find equivalent fractions.	23, 24	12-1, 12-2, 47-2
	C-13	Compare 2 proper fractions and order 5 proper fractions.	25	13-1
	C-14	Interchange mixed numbers and improper fractions.	22	14-1
		Fraction Operations		
	C-15	Add/subtract fractions with common denominators.	27	15-1
	C-16	Add/subtract mixed numbers with common denominators.	28, 29	16-1, 16-2
	C-17	Add/subtract unlike proper fractions.	30, 31	17-1, 17-2, 45-3
	C-18	Find common denominators.	30, 31	17-1, 17-2, 18-1
	C-19	Multiply 2 proper fractions or a proper fraction by a whole number.	32, 33, 48	19-1, 19-2, 45-3
	C-20	Divide proper fractions by proper fractions or whole numbers.	34	20-1, 45-3
		Decimal Concepts		
	C-21	Write decimals from a picture or from a number line.	35, 36	21-1
	C-22	Read and write decimals up to thousandths.	36	22-1
	C-23	Identify place value up to ten-thousandths.	37	23-1, 23-2
	C-24	Compare/order decimals up to hundredths.	38, 39	24-1
	C-25	Interchange fractions having denominators of 10 or 100 with decimals.	35, 40	25-1

Teacher Manual Foreword

		Decimal Operations	Student Book	Skill Builders
	C-26	Add and subtract decimals or money.	41, 42, 43	26-1
	C-27	Multiply money and 2-place decimals.	44, 45	27-1, 27-2
	C-28	Divide money and up to 2-place decimals.	46, 47, 49	28-1, 28-2, 45-4
	C-29	Identify the percent of shaded figures.		29-1
	C-30	Interchange 2-place decimals with fractions.		30-1
		Geometry		
	C-31	Identify a point, line, line segment, ray and angle.		31-1
	C-32	Identify lines.	50	32-1
	C-33	Identify angles.	51, 52	33-1
Test Item/Objective Number	C-34	Identify basic shapes and solids.	53	34-1
	C-35	Identify parts of a circle.	54	35-1
		Measurement		
	C-36	Measure to the nearest 1/8 unit. Convert measurements of length.	55	36-1, 36-2
	C-37	Use a protractor to measure and draw angles.	51	37-1
	C-38	Find the perimeter or area of polygons.	56, 57	38-1, 38-2
	C-39	Find the volume of a rectangular solid.	58	39-1, 39-2
	C-40	Tell time to the nearest minute and find intervals of time.	59	40-1
	C-41	Use the appropriate unit for weight and convert measurements.	60	41-1
	C-42	Use the appropriate unit for liquid capacity and convert measurements.		42-1
	C-43	Give the total value of a set of coins and bills; make change for a $20 bill.		43-1
		Problem Solving		
	C-44	Find the missing number in patterns, coordinate grids.	2, 61	44-1
	C-45	Solve 1-step or 2-step word problem with whole numbers.	16, 17, 48, 49	45-1, to 45-5
	C-46	Find the average of whole numbers or decimals.	18	46-1, 46-2
	C-47	Read and interpret pictographs, bar graphs, tables and charts, probability.	19, 61, 62, 64	47-1, 47-2
	C-48	Read and interpret line graphs and circle graphs.	63	48-1
	C-49	Estimate sums and differences of numbers up to and including 4 digits.	7, 8, 16, 26	49-1, 49-2
	C-50	Estimate products of a 3-digit number.	14, 15, 17	50-1, 50-2

Grade 6 Supply List

The following lists summarize materials suggested in the Grade 6 Extensions Teacher Manual.

Teachers should have the listed **Classroom Materials** on hand. Teachers should also have at least one **Manipulative Kit** (or the equivalent manipulatives already in the classroom.) **Other Materials** may be found in the home or the classroom. The **Overhead Manipulative Kit** is optional. All optional materials are indicated with a ▶.

Classroom Materials

index cards
posterboard/chart paper
crayons
string
glue
rulers - standard and metric
scissors
tape - masking and clear
▶ tagboard
▶ construction paper
▶ colored chalk
▶ primer balance
▶ calculators

Grade 6 Manipulative Kit

1	Base Ten Grade 6 Set: 200 ones, 100 tens, 20 hundreds, 2 thousands
200	interlocking cubes
5	place value mats Fraction Bars®
5	geoboards with geobands
5	10-sided dice
8	6-sided dice

Other Materials

real or play coins and bills
playing cards
empty egg cartons
paper plates
game markers (corn, beans, etc.)
small box of animal crackers
2 pounds of butter in 1/4 lb. sticks
tennis balls or soft rubber balls
clock with moveable hands
1-pound can of coffee
▶ stick of gum
▶ square tiles
▶ tortilla rounds
▶ plastic knives
▶ needlepoint canvas squares
▶ local fast food menu
▶ order pads
▶ 2 boxes of cereal
▶ compasses/chalkboard compass
▶ protractor/chalkboard protractor
▶ toothpicks or straws
▶ apples or cardboard circles
▶ chocolate bars scored into 12ths

Overhead Manipulative Kit

▶ base ten block pieces
▶ colored squares
▶ Fraction Bars®
▶ geoboards
▶ coins and bills
▶ playing cards

Teacher Manual Foreword

Lesson Pacing Plans

Lesson plans provide the teacher with a guide to what they should complete each lesson.

The pacing plan on the following 4 pages covers the *Extensions* materials in 20 lessons. Teachers with limited time may want to focus on the sections that are most important to their class or school district. Teachers with more time may do fewer pages per lesson than suggested in this calendar.

Explanation of the first lesson of the pacing calendar:

	Lesson 1
Warm-up	Daily Review #1 (in back of student book)
Hands On Lesson	**Objective:** Discover the pattern, comparing numbers, rounding patterns, commutative property **Materials:** Base ten blocks, Place Value Mats, dice, cards, egg cartons, Masters 2, 7 **Teacher Guide pages:** 1-4
Student Practice	Student book pages 1-4
Journal Prompt/ Performance Assessment	Teacher Guide page 3
Second Warm-up Reteaching (Homework)	Daily Review #2 (in back of student book) Skill Builders 2-1 (in back section of this manual)
Games	Place Value Bingo Comparison Make the Greatest Number Rounding Relays

- These quick, 5-question reviews are found in back of the student book. Students record results on the inside cover of their books.
- This section shows the goals for the lesson, the materials needed, and the lesson plan pages.
- These are the pages students will complete in their books. To save time, only assign even problems.
- This section lists the performance assessment for the lesson. Use the blank Journal Prompt forms in the Assessment section as a copy template.
- Use these pages as an integrated review-and-reteach system to prepare students for tests.
- Easy-to-complete games keep students engaged.

20-Lesson Pacing Plan for Grade 6

	Lesson 1	Lesson 2	Lesson 3	Lesson 4	Lesson 5
Warm-up	Daily Review #1 (in back of student book)	Test Day	Daily Review #3 (in back of student book)	Daily Review #5 (in back of student book)	Daily Review #7 (in back of student book)
Hands On Lesson	**Objective:** Discover the pattern, comparing numbers, rounding patterns, commutative property **Materials:** Base ten blocks, Place Value Mats, dice, cards, egg cartons, Masters 2, 7 **Teacher Guide pages: 1-4**	Administer the Pre-Test. Record results on Class Record Sheet and Student Progress Report (in Test Assessment Pack). Transfer results to pg. 3 of the Parent Handbook and send home with student.	**Objective:** Addition and subtraction, estimating sums, and differences **Materials:** Base ten blocks, Place Value Mats, dice, cards, Masters 2, 3, 4, calculators (optional) **Teacher Guide pages: 5-8**	**Objective:** Multiplying and dividing by 1- and 2-digit numbers **Materials:** Base ten blocks, Place Value Mats, index cards, dice, tennis balls, paper plates, Masters 5, 6, calculators (optional) **Teacher Guide pages: 9-12**	**Objective:** Estimating products and quotients, steps in problem solving **Materials:** Base ten blocks, paper plates, cards, Place Value Mats, index cards, chart or poster paper **Teacher Guide pages: 13-16**
Student Practice	Student book pages 1-4	Test Day	Student book pages 5-8	Student book pages 9-12	Student book pages 13-16
Journal Prompt/ Performance Assessment	Teacher Guide page 3	Optional Journal Prompt on Teacher Guide page 4	Teacher Guide page 7	Teacher Guide page 10	Teacher Guide page 14
Second Warm-up Reteaching (Homework)	Daily Review #2 (in back of student book) Skill Builders 2-1 (in back section of this manual)		Daily Review #4 (in back of student book) Skill Builders 49-1 or 49-2 (in back section of this manual)	Daily Review #6 (in back of student book) Skill Builders 10-1 (in back section of this manual)	**Problem Solving:** Skill Builders 45-5
Games	Place Value Bingo Comparison Make the Greatest Number Rounding Relays		Estimating Sums Estimating Sums and Differences Relays	Opposite Ball Smallest Quotient	Estimating Products Relay

© Math Teachers Press, Inc.

20-Lesson Pacing Plan for Grade 6

	Lesson 6	Lesson 7	Lesson 8	Lesson 9	Lesson 10
Warm-up	Daily Review #8 (in back of student book)	Daily Review #10 (in back of student book)	Daily Review #12 (in back of student book)	Daily Review #14 (in back of student book)	Daily Review #16 (in back of student book)
Hands On Lesson	**Objective:** Problem solving strategies, average, collecting data **Materials:** Cubes, paper plates, dice, chart or poster paper, index cards, calculator (optional), counters, animal crackers **Teacher Guide pages:** 17-19	**Objective:** Fraction concepts and terms **Materials:** Fraction Bars, 2 pounds of butter, bills and coins (or Master 8), apples or cardboard circles (optional) **Teacher Guide pages:** 20-22	**Objective:** Equivalent fractions, comparing fractions **Materials:** Fraction Bars, chocolate bars scored into 12 pieces, cubes, square tiles **Teacher Guide pages:** 23-25	**Objective:** Addition and subtraction of like fractions and mixed numbers, estimation **Materials:** Fraction Bars, tortilla rounds, plastic knives **Teacher Guide pages:** 26-28	**Objective:** Addition and subtraction of unlike fractions **Materials:** Fraction Bars, paper plates, dice, multiple strips (made from Master 7), cubes **Teacher Guide pages:** 29-31
Student Practice	Student book pages 17-19	Student book pages 20-22	Student book pages 23-25	Student book pages 26-28	Student book pages 29-31
Journal Prompt/ Performance Assessment	Teacher Guide page 18	Teacher Guide page 21	Teacher Guide page 23	Teacher Guide page 26	Teacher Guide page 31
Second Warm-up Reteaching (Homework)	Daily Review #9 (in back of student book) Skill Builders 46-1 (in back section of this manual)	Daily Review #11 (in back of student book) Skill Builders 11-1 (in back section of this manual)	Daily Review #13 (in back of student book) Skill Builders 12-1 (in back section of this manual)	Daily Review #15 (in back of student book) Skill Builders 15-1 (in back section of this manual)	**Problem Solving:** Skill Builders 45-1
Games	Averaging Dice	What's My Secret?	Simplest Fraction Bingo (Skill Builders 12-2)	Make the Greatest Difference	Dicey Differences

Teacher Manual Foreword © Math Teachers Press, Inc.

20-Lesson Pacing Plan for Grade 6

	Lesson 11	**Lesson 12**	**Lesson 13**	**Lesson 14**	**Lesson 15**
Warm-up	Daily Review #17 (in back of student book)	Daily Review #19 (in back of student book)	Daily Review #21 (in back of student book)	Daily Review #23 (in back of student book)	Daily Review #25 (in back of student book)
Hands On Lesson	**Objective:** Multiplying and dividing fractions **Materials:** Egg cartons, cubes, Fraction Bars, dice, paper, crayons **Teacher Guide pages:** 32-34	**Objective:** Relating fractions and decimals, decimal place values, comparing and ordering decimals **Materials:** Fraction Bars, base ten blocks, Place Value Mats, bills and coins (or Master 8), Master 2, index cards, Master 1, game markers **Teacher Guide pages:** 35-38	**Objective:** Equivalent decimals, relating fractions and decimals, adding and subtracting decimals **Materials:** Base ten blocks, dice, coins and bills (or Master 8), Fraction Bars, Masters 10 and 11 (for decimal place value mats) **Teacher Guide pages:** 39-42	**Objective:** Multiplying and dividing decimals, reading tables **Materials:** Base ten blocks, coins and bills (or Master 8), dice, paper plates, local fast food menu, Place Value Mats, order pads, centimeter squares (made from Master 2) **Teacher Guide pages:** 43-46	**Objective:** Dividing decimals, fractional part of a number, word problems **Materials:** Base ten blocks, paper plates, coins and bills, 2 boxes of cereal, string, counters **Teacher Guide pages:** 47-49
Student Practice	Student book pages 32-34	Student book pages 35-38	Student book pages 39-42	Student book pages 43-46	Student book pages 47-49
Journal Prompt/ Performance Assessment	Teacher Guide page 32	Teacher Guide page 38	Teacher Guide page 39	Teacher Guide page 45	Teacher Guide page 49
Second Warm-up Reteaching (Homework)	Daily Review #18 (in back of student book) Skill Builders 19-1 or 19-2 (in back section of this manual)	Daily Review #20 (in back of student book) Skill Builders 23-1 (in back section of this manual)	Daily Review #22 (in back of student book) Skill Builders 25-1 (in back section of this manual)	Daily Review #24 (in back of student book) Skill Builders 27-2 (in back section of this manual)	**Problem Solving:** Skill Builders 45-5
Games	Dicey Fraction War	Place Value Bingo Fraction War	Play previous games.	Play previous games.	Play previous games.

© Math Teachers Press, Inc.   Teacher Manual Foreword

20-Lesson Pacing Plan for Grade 6

	Lesson 16	Lesson 17	Lesson 18	Lesson 19	Lesson 20
Warm-up	Daily Review #26 (in back of student book)	Daily Review #28 (in back of student book)	Daily Review #30 (in back of student book)	Test Day	Daily Review #32 (in back of student book)
Hands On Lesson	**Objective:** Geometry terms, measuring angles **Materials:** Geoboards, protractors (or Master 9), toothpicks or straws, glue, Master 12, masking tape, scissors, tagboard (optional) **Reminder:** Tell students to return completed Parent Handbook to receive Certificate of Achievement. **Teacher Guide pages:** 50-53	**Objective:** Parts of a circle, making and using a ruler, perimeter, area **Materials:** Centimeter ruler (or Master 9), circular geoboard (or Master 12), units blocks, Master 2, Skill Builders 36-1, masking tape, compasses/chalkboard compass **Teacher Guide pages:** 54-57	**Objective:** Volume, intervals of time, estimating and converting weight, coordinate grids **Materials:** Units blocks, 2 clocks with moveable hands, geoboards, Master 12, Master 2, 1 lb. Coffee can, scissors, tape, stick of gum, primer balance (optional) **Teacher Guide pages:** 58-61	Administer the Post-Test. Record results on Class Record Sheet and Student Progress Report in Assessment Test Pack.	**Objective:** Collecting data, bar graphs, line graphs, tables **Materials:** Fast food menu, index cards, one inch graph paper (Master 7), order pads, coins and bills (or Master 8), calculators (optional) **Teacher Guide pages:** 62-64
Student Practice	Student book pages 50-53	Student book pages 54-57	Student book pages 58-61		Student book pages 62-64
Journal Prompt/ Performance Assessment	Teacher Guide page 52	Teacher Guide page 56	Teacher Guide page 58	Select a problem from page 16 or 17 or create an appropriate problem for the final journal prompt (on the last page of the Post-Test).	Teacher Guide page 63
Second Warm-up Reteaching (Homework)	Daily Review #27 (in back of student book) Skill Builders 32-1 or 33-1 (in back section of this manual)	Daily Review #29 (in back of student book) Skill Builders 38-1 or 38-2 (in back section of this manual)	Daily Review #31 (in back of student book) Skill Builders 39-1 (in back section of this manual)		**Problem Solving:** Skill Builders 45-3
Games		Play previous games.			Probability (Skill Builders 47-2)

Blank Pacing Plan

	Lesson #	Lesson #	Lesson #	Lesson #	Lesson #
Warm-up					
Hands On Lesson					
Student Practice					
Journal Prompt Performance Assessment					
Second Warm-up Reteaching (Homework)					
Games					

Assessment

This section contains:

Student Progress Report 1
Class Record Sheet 2
Pre-Test. 5
Post-Test 11
Final Journal Prompt (optional). 17
Answer Keys 18
Glossary 25
Journal Prompt Instructions and Scoring Guide. 29
Blank Math Journal. 31

The Test Assessment Pack includes Pre-Tests, Post-Tests, Student Progress Report and a Class Record Sheet. Copies of these same tests and record sheets may be found on pages 1–16 of this Assessment section for teachers who wish to make copies rather than use the Test Assessment Pack. Instructions for teachers who will make copies from this section are given in parentheses after each step.

Follow these steps to administer the Pre- and Post-Test:

1. Distribute a copy of the **Pre-Test** from the Test Assessment Pack to each student.
   (Or make copies of the Pre-Test from pages 5-10.)

2. Administer the **Pre-Test** and record the results on the **Class Record Sheet** found in the Test Assessment Pack. Record missed objectives in the first row of boxes for each student. If a column is shaded, the objective is not tested in this grade level.
   (Or make a copy of the **Class Record Sheet, on pages 2 and 3** and tape the two sides together.)

   Results may also be recorded on the **Student Progress Report** (last page of the Pre-Test in the Test Assessment Pack.) This report may serve as a report card for each student. (Or copy page 1 from the assessment section.)

3. Administer the **Post-Test** from the Test Assessment Pack at the end of the session. Record results on the **Class Record Sheet** and **Student Progress Report**.
   (Or make copies of the Post-Test from pages 11–16.)

4. If journal prompts have been used during the session, teachers may wish to include a final journal prompt on the last page of the Post-Test. The teacher should suggest a word problem of appropriate difficulty for the class, similar to those taught on page 16 or 17 of the Student Book.

Student Progress Report

Grade 6

Mark an X in the Pre- and/or Post-Test boxes to indicate <u>missed objectives</u>.
Record the number correct out of 50 possible.

Teacher _____

Student _____ School _____

Pre-Test Post-Test

Numeration
- ☐ ☐ **C-1** Identify the place value in a 7-digit number.
- ☐ ☐ **C-2** Read, write and compare 9-digit numbers.
- ☐ ☐ **C-3** Round to the nearest thousand.
- ☐ ☐ **C-4** Identify prime numbers and the factors of composite numbers up to 100.
- ☐ ☐ **C-5** Use the commutative, associative or the distributive property.

Whole Number Operations
- ☐ ☐ **C-6** Add numbers up to 6-digits.
- ☐ ☐ **C-7** Subtract numbers up to 6-digits.
- ☐ ☐ **C-8** Multiply a 3-digit number by a 2-digit number. Multiply by multiples of 10.
- ☐ ☐ **C-9** Divide a 6-digit by a 1-digit number.
- ☐ ☐ **C-10** Divide a 4-digit by a 2-digit number.

Fractions
- ☐ ☐ **C-11** Write fractions from shaded regions, number lines and printed words.
- ☐ ☐ **C-12** Find equivalent fractions.
- ☐ ☐ **C-13** Compare 2 like or unlike proper fractions and order 5 like or unlike proper fractions.
- ☐ ☐ **C-14** Interchange mixed numbers and improper fractions.
- ☐ ☐ **C-15** Add/subtract fractions with common denominators.
- ☐ ☐ **C-16** Add/subtract mixed numbers with common denominators.
- ☐ ☐ **C-17** Add/subtract unlike proper fractions.
- ☐ ☐ **C-18** Add/subtract unlike mixed numbers.
- ☐ ☐ **C-19** Multiply 2 proper non-reducible fractions or a proper fraction by a whole number.
- ☐ ☐ **C-20** Divide proper fractions by proper fractions or whole numbers.

Decimals
- ☐ ☐ **C-21** Write decimals from a picture or from a number line.
- ☐ ☐ **C-22** Read and write decimals up to thousandths.
- ☐ ☐ **C-23** Identify place value up to ten-thousandths.
- ☐ ☐ **C-24** Compare/order decimals up to hundredths.
- ☐ ☐ **C-25** Interchange fractions having denominators of 10 or 100 with decimals.

Pre-Test Post-Test

- ☐ ☐ **C-26** Add and subtract decimals or money.
- ☐ ☐ **C-27** Multiply money and 2-place decimals.
- ☐ ☐ **C-28** Divide money and up to 2 place decimals.
- ☐ ☐ **C-29** Identify the percent of shaded figures.
- ☐ ☐ **C-30** Interchange 2-place decimals with fractions.

Geometry & Measurement
- ☐ ☐ **C-31** Identify a point, line, line segment, ray and angle.
- ☐ ☐ **C-32** Identify lines.
- ☐ ☐ **C-33** Identify angles.
- ☐ ☐ **C-34** Identify basic shapes and solids.
- ☐ ☐ **C-35** Identify parts of a circle.
- ☐ ☐ **C-36** Measure to the nearest $\frac{1}{8}$ unit.
- ☐ ☐ **C-37** Use a protractor to measure and draw angles.
- ☐ ☐ **C-38** Find the perimeter or area.
- ☐ ☐ **C-39** Find the volume of a rectangular solid.
- ☐ ☐ **C-40** Tell time to the nearest minute.
- ☐ ☐ **C-41** Use the appropriate unit for weight.
- ☐ ☐ **C-42** Use the appropriate unit for liquid capacity.
- ☐ ☐ **C-43** Give the total value of a combination of coins and bills; make change for a $20 bill.

Problem Solving
- ☐ ☐ **C-44** Can find the missing number in patterns.
- ☐ ☐ **C-45** Can solve a 1-step word problem with whole numbers.
- ☐ ☐ **C-46** Find the average of whole numbers or decimals.
- ☐ ☐ **C-47** Read and interpret pictographs, bar graphs, tables and charts.
- ☐ ☐ **C-48** Read and interpret line graphs and circle graphs.
- ☐ ☐ **C-49** Estimate sums and differences of numbers up to and including 4 digits.
- ☐ ☐ **C-50** Estimate products of a 3-digit number.

☐/50 ☐/50 **Total Scores (out of 50 possible)**

Assessment 1

Grade 6 Class Record Sheet

Teacher: _____

Student Name:	State Standard	1	2	3	4	5	6	7	8	9	10	11	12	13	14	15	16	17	18	19	20	21	22	23	24	25
Problem # / Strand		\multicolumn{5}{c}{Numeration}	\multicolumn{5}{c}{Operations}	\multicolumn{10}{c}{Fractions}	\multicolumn{5}{c}{Decimals}
1. Pre- / Post-																										
2. Pre- / Post-																										
3. Pre- / Post-																										
4. Pre- / Post-																										
5. Pre- / Post-																										
6. Pre- / Post-																										
7. Pre- / Post-																										
8. Pre- / Post-																										
9. Pre- / Post-																										
10. Pre- / Post-																										
11. Pre- / Post-																										
12. Pre- / Post-																										
13. Pre- / Post-																										
14. Pre- / Post-																										
15. Pre- / Post-																										
16. Pre- / Post-																										
17. Pre- / Post-																										
18. Pre- / Post-																										
19. Pre- / Post-																										
20. Pre- / Post-																										
% of students with correct answers Pre- / Post-																										

Assessment

Class/School: _____

- Put an X on missed objectives.
- Record days absent in the last column.

Grade 6

																									No. correct out of 50	% correct	Abs.
26	27	28	29	30	31	32	33	34	35	36	37	38	39	40	41	42	43	44	45	46	47	48	49	50			
Decimals (cont.)					**Geometry**					**Measurement**								**Mixed**									
																									/50		
																									/50		
																									/50		
																									/50		
																									/50		
																									/50		
																									/50		
																									/50		
																									/50		
																									/50		
																									/50		
																									/50		
																									/50		
																									/50		
																									/50		
																									/50		
																									/50		
																									/50		
																									/50		
																									/50		
																									/50		
																									/50		
																									/50		
																									/50		
																									/50		
																									/50		
																									/50		
																									/50		
																									/50		
																									/50		
																									/50		
																									/50		
																									/50		
																									/50		
																									Mean %, Pre-		
																									Mean %, Post-		

Assessment 3

Grade 6 Pre-Test

1. What digit is in the ten millions place in the number 257,314,698?

   _____

2. Write seven million three hundred sixty-five thousand as a numeral:

   _____

3. A stadium sold 27,365 tickets for Thursday night's football game. What is this number rounded to the nearest thousand?

   _____

4. Which of the following is a prime number?

   | 5 | 6 | 8 | 9 |

   _____

5. What number goes in the box?

   _____

   $5 \times (1 + 3) = (5 \times \boxed{}) + (5 \times 3)$

6.   2,423
      1,316
      3,275
   + 1,102

   _____

7. Hiking trail A is 6,427 feet long. Hiking trail B is 8,201 feet long. How much longer is trail B?

   _____

8. A movie theater sells 741 tickets each day. How many tickets are sold in 46 days?

   _____

9. $6 \overline{)1812}$

   _____

10. A factory shipped 3,213 jars in 51 boxes. How many jars in each box?

    _____

 Teacher Note: You may help students read words when requested. Do not explain the meaning of the words.

Assessment

11.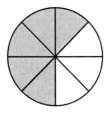

    What fraction is shaded? _____

12. Simplify $\frac{6}{15}$. _____

13. Select the set of the fractions ordered from least to greatest.

    A $\frac{1}{8}$ $\frac{1}{4}$ $\frac{1}{2}$ $\frac{3}{8}$ $\frac{3}{4}$

    B $\frac{1}{2}$ $\frac{1}{4}$ $\frac{3}{4}$ $\frac{1}{8}$ $\frac{3}{8}$

    C $\frac{1}{8}$ $\frac{3}{8}$ $\frac{1}{4}$ $\frac{1}{2}$ $\frac{3}{4}$

    D $\frac{1}{8}$ $\frac{1}{4}$ $\frac{3}{8}$ $\frac{1}{2}$ $\frac{3}{4}$

14. Change $\frac{14}{3}$ to a mixed number. _____

15. Juanita bought $\frac{1}{8}$ pound of taffy and $\frac{5}{8}$ pound of jelly beans. How much candy did Juanita buy? (Express your answer in simplest form.) _____

16. Subtract and simplify. _____

    $$6\frac{1}{5}$$
    $$-2\frac{3}{5}$$

17. Tyrell used $\frac{2}{3}$ cup of detergent for a large load of laundry and $\frac{1}{4}$ cup for a small load. How much detergent did he use? _____

18. Subtract and simplify. _____

    $$3\frac{2}{3}$$
    $$-1\frac{1}{6}$$

19. Jane had $\frac{3}{4}$ pound of butter. She wanted to use $\frac{2}{3}$ of it for a recipe. How much butter did she use? (in lowest terms) _____

20. $\frac{2}{5} \div \frac{3}{4} =$ _____

21. What decimal is shown at Point A on the number line? _____

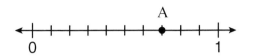

22. Write twenty-four hundredths as a decimal numeral: _____

23. What is the place name of the underlined digit in 0.7654? _____

24. Which number sentence is true? _____

    A 0.23 > 0.20

    B 0.23 < 0.20

    C 0.23 = 0.20

    D 0.20 > 0.23

25. Change $\frac{37}{100}$ to a decimal. _____

26. Tim's frog jumped 5.348 ft. Ryan's frog jumped 2.152 ft. How much farther did Tim's frog jump? _____

27. 0.21
    × 0.32
    _____

28. Darnell has $7.26. He wants to buy pencils that cost $.06 each. How many pencils can Darnell buy? _____

29.

What percent of the square is shaded? _____

30. Change $\frac{21}{100}$ to a percent. _____

6E Pre-Test – Page 4

31. Which of the following shows a line?

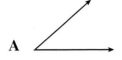

32. Which lines are perpendicular?

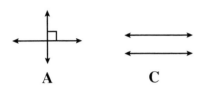

33. Angle A is what kind of angle?

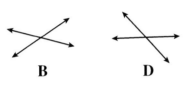

- **A** obtuse
- **B** right
- **C** acute
- **D** straight

34. Which is a parallelogram?

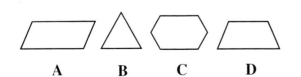

35.

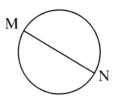

$\overline{MN}$ is what part of the circle?

- **A** center
- **B** circumference
- **C** diameter
- **D** radius

36. What is the length of $\overline{AB}$ to the nearest $\frac{1}{4}$ inch?

_____ in.

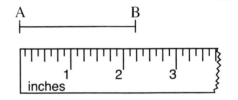

- **A** 2 inches
- **B** $2\frac{1}{8}$ inches
- **C** $2\frac{1}{4}$ inches
- **D** $2\frac{3}{8}$ inches

37.

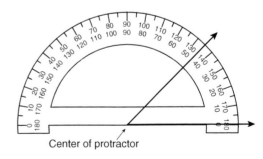

What is the measure of the angle?

38.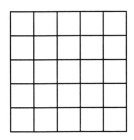

If ☐ = 1 square unit, the area of the figure below is how many square units?

_____

39.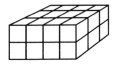

☐ = 1 cubic unit. What is the volume in cubic units?

_____

40. Football practice starts at 4:30 and lasts for 2 hours, 15 minutes. What time will practice end?

_____

41. What is the approximate weight of a book?

A 1 gram
B 1 kilogram
C 1 ton
D 1 ounce

42.
```
1 pint = 2 cups
1 quart = 2 pints
1 gallon = 4 quarts
```

3 quarts = _____ cups

43. How much money?

_____

44. What is the missing number in the pattern?

_____

36, 32, ☐, 24, 20

45. A $12.00 book is marked 1/4 off. What is the amount of the discount? _____

46. The heights of the five people on the basketball team are 44 inches, 50 inches, 52 inches, 48 inches and 46 inches. What is the average height? _____

47.

Name	Miles
Jon	$8\frac{1}{2}$
Betsy	9
Ivan	8
Sia	$10\frac{1}{2}$

How many more miles did Betsy jog than Jon? _____

48.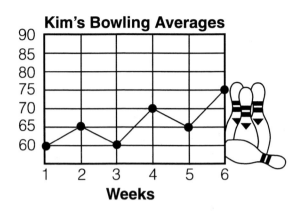

How many points did Kim's bowling average improve from week 1 to week 4? _____

49. Estimate by rounding.
384 + 517 is about:

A 700
B 800
C 900
D 1,000

50. Estimate by rounding.
395 × 72 is about:

A 2,100
B 2,800
C 21,000
D 28,000

Name _____

Grade 6 Post-Test

1. What digit is in the hundred thousands place in the number 257,314,698?

   _____

2. Write the number 328,000,000 in words.

   _____
   _____

3. A lake measured 42,810 feet long. What is the length rounded to the nearest thousand?

   _____

   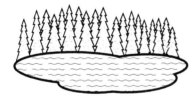

4. Which of the following is a prime number?

       3   4   6   9

   _____

5. What number goes in the box?

   $3 \times (2 + 4) = (3 \times \boxed{}) + (3 \times 4)$

   _____

6.   4,012
      2,517
      3,023
   + 1,490

   _____

7. 9,203 fans went to a baseball game. 3,564 of them left before the fireworks show. How many stayed for the show?

   _____

8. The health club has 741 members. Membership fees are $64 per month for each member. How much does the club make on membership each month?

   _____

9. $9\overline{)3717}$

   _____

10. A store sold 43 bottles of perfume for a total of $1,892. What was the price of one bottle of perfume?

    _____

Assessment   11

6E Post-Test – Page 2

11.

    What fraction is shaded? _____

12. Simplify $\frac{18}{24}$. _____

13. Select the set of the fractions ordered from least to greatest.

    A $\frac{1}{10}$ $\frac{3}{10}$ $\frac{1}{2}$ $\frac{1}{5}$ $\frac{2}{5}$

    B $\frac{1}{2}$ $\frac{1}{5}$ $\frac{1}{10}$ $\frac{2}{5}$ $\frac{3}{5}$

    C $\frac{1}{10}$ $\frac{3}{10}$ $\frac{1}{5}$ $\frac{1}{2}$ $\frac{2}{5}$

    D $\frac{1}{10}$ $\frac{1}{5}$ $\frac{3}{10}$ $\frac{2}{5}$ $\frac{1}{2}$

14. Change $\frac{17}{5}$ to a mixed number. _____

15. Pete had $\frac{7}{8}$ foot of framing material. He used $\frac{3}{8}$ foot. How much framing material does he have left? (Express your answer in simplest terms.) _____

16. Add and simplify.

    $2\frac{4}{5}$
    $+ 1\frac{3}{5}$
    _____

17. Samantha ran $\frac{7}{10}$ of a mile. Then she walked $\frac{2}{5}$ of a mile. How much farther did she run than walk?

    _____

18. Add and simplify.

    $3\frac{2}{5}$
    $+ 4\frac{1}{2}$
    _____

19. Bonnie had $\frac{3}{5}$ gallon of paint. She used $\frac{2}{3}$ of the paint to finish a wall. How much paint did she use? (Express your answer in simplest terms.)

    _____

20. $\frac{3}{8} \div \frac{2}{5} =$ _____

6E Post-Test – Page 3

21. Which point on the number line represents 0.4?

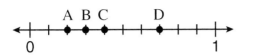

22. The correct words for 0.73 are:

    A   seventy-three hundred
    B   seventy-three tenths
    C   seventy-three thousandths
    D   seventy-three hundredths

23. What digit is in the thousandths place in the number 0.5139?

24. Which number sentence is true?

    A. 0.62 = 0.60
    B. 0.60 > 0.62
    C. 0.62 > 0.60
    D. 0.62 < 0.60

25. Change $\frac{51}{100}$ to a decimal.

26. Jane ran 3.847 miles on Saturday and 2.365 miles on Sunday. How much farther did she run on Saturday?

27. 0.23
    × 0.31
    ─────

28. Paula bought 6.5 pounds of hamburger for a cookout. If she uses 0.5 pound for each burger, how many burgers can she make?

29.

What percent of the square is shaded?

30. Change $\frac{17}{100}$ to a percent.

31. Which of the following shows a line segment?

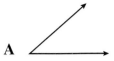

A
B
C
D

32. Which lines are perpendicular?

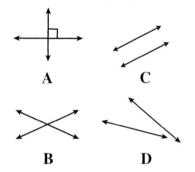

A
B
C
D

33.

What kind of angle is ∠XYZ?

A acute
B right
C straight
D obtuse

34. Which is a hexagon?

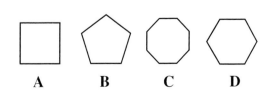

A   B   C   D

35.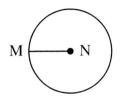

$\overline{MN}$ is what part of the circle?

A center
B circumference
C diameter
D radius

36. What is the measure of the nail to the nearest $\frac{1}{4}$ inch?

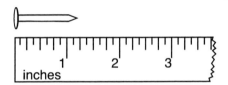

A 1 inch   C $1\frac{3}{8}$ inches

B $1\frac{1}{4}$ inch   D $1\frac{1}{2}$ inches

37.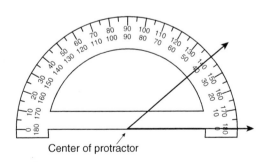

Center of protractor

What is the measure of the angle?

38.

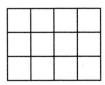

If ▢ = 1 square unit, the area of the figure below is how many square units?

_____

39.

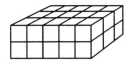

▢ = 1 cubic centimeter. What is the volume in cubic centimeters?

_____

40. What time will it be 2 hours and 20 minutes after 6:15?

_____

41. What is the approximate weight of a paper clip?

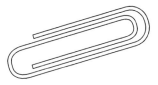

A   1 gram
B   1 pound
C   1 kilogram
D   1 ton

42.

> 1 pint = 2 cups
> 1 quart = 2 pints
> 1 gallon = 4 quarts

2 gallons = _____ pints

43. How much money?

_____

44. What is the missing number in the pattern?

_____

29, 25, ▢, 17, 13

45. An $8.00 record was marked $\frac{1}{4}$ off. What is the amount of discount?

_____

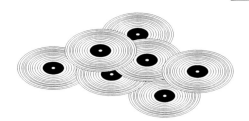

46. In 4 games, Jane scored 12, 13, 16, and 15 points. What is the average number of points scored in each game?

_____

47.

Tony's Allowance	
Snacks	$2.50
Savings	$3.00
Movie	$5.50
Clothes	$4.00

How much more money does Tony spend on movies than on snacks?

_____

48.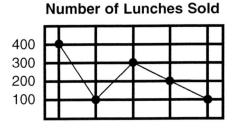

Number of Lunches Sold

How many more lunches were sold on Monday than on Tuesday?

_____

49. Estimate by rounding.
718 − 296 is about:

A 300
B 400
C 500
D 600

_____

50. Estimate by rounding.
610 × 79 is about:

A 4,200
B 4,800
C 42,000
D 48,000

_____

Final Journal Prompt

Five Steps in Problem Solving:

Step 1: Read and understand.
Step 2: Find the question and needed facts.
Step 3: Decide on a process.
Step 4: Estimate.
Step 5: Solve and check back.

Explain how you would use the five steps in problem solving to solve this problem. Remember to show your work and draw pictures to help explain your answer.

Answer:

 Teacher Note: Select a word problem similar to those taught on Student Book page 13 or 26 for a Journal Prompt.

Pre- & Post Tests, Daily Reviews — Answer Keys

Pre-Test
1. 5 2. 7,365,000 3. 27,000 4. 5 5. 1 6. 8116
7. 1774 8. 34,086 9. 302 10. 63 11. ⁵/₈ 12. ²/₅
13. D 14. 4 ²/₃ 15. ³/₄ lb. 16. 3 ³/₅ 17. ¹¹/₁₂
18. 2 ½ 19. ½ lb. 20. ⁸/₁₅ 21. 0.7 22. 0.24
23. tenths 24. A 25. 0.37 26. 3.196 27. 0.0672
28. 121 29. 25% 30. 21% 31. D 32. A
33. C acute 34. A 35. C diameter 36. C
37. 45° 38. 25 39. 24 cu. units 40. 6:45 41. B
42. 12 cups 43. $7.88 44. 28 45. $3.00
46. 48 in. 47. ½ mile 48. 10 49. C 900
50. D 28,000

Post-Test
1. 3 2. three hundred twenty-eight million
3. 43,000 4. 3 5. 2 6. 11,042 7. 5,639
8. $47,424 9. 413 10. $44 11. ³/₅ 12. ³/₄
13. D 14. 3 ²/₅ 15. ½ ft. 16. 4 ²/₅ 17. ³/₁₀
18. 7 ⁹/₁₀ 19. ²/₅ gal. 20. ¹⁵/₁₆ 21. Point C 22. D
23. 3 24. C 25. 0.51 26. 1.482 27. 0.0713
28. 13 29. 20% 30. 17% 31. D 32. A
33. B right 34. D 35. D radius 36. B 37. 40°
38. 12 39. 30 cu. cm 40. 8:35 41. A 1 gram
42. 16 43. $3.94 44. 21 45. $2.00 46. 14
47. $3.00 48. 300 49. B 400 50. D 48,000

Daily Review 1
1. (one) billions 2. six hundred fifty billion 3. 12,000 4. (D) 14 5. 5

Daily Review 2
1. 81,721 2. 8118 3. 75,110 4. 6589 R2 5. 205

Daily Review 3
1. (D) ⁷/₃ 2. 8 3. ¹/₅ ²/₅ ³/₅ ⁴/₅ ⁵/₅ 4. 2½ 5. 7

Daily Review 4
1. ³/₇ 2. 5 ³/₄ 3. 4 ⅛ 4. ¹⁵/₃₂ 5. ⅙

Daily Review 5
1. B 2. 0.325 3. 1 4. 2 ³/₁₀ 5. ⁵/₁₂

Daily Review 6
1. 0.18 2. 1.37 3. 10% 4. 61% 5. $16.35

Daily Review 7
1. 6 2. 0.966 3. 3.2 4. 75%
5. (C) diameter

Daily Review 8
1. (C) angle 2. (B) 3. (C) right 4. (C)
5. .73

Daily Review 9
1. hundredths 2. 2.565 3. 76.1 4. 33% 5. 4 ⅛

Daily Review 10
1. = 2. (A) acute 3. (D) 4. (C) meter 5. 3%

Daily Review 11
1. (C) 1 ³/₈ 2. (C) pounds 3. $5.10 4. 149 mi
5. daylight

Daily Review 12
1. 24 2. (D) 130° 3. 40 sq. in. 4. 24 cu. cm.
5. 5 c.

Daily Review 13
1. (B) 35 2. (B) 15° 3. 1.9 mi. 4. 14¢
5. 6:37

Daily Review 14
1. (B) gram 2. 4500 3. $11.50 4. 2400 cu. ft.
5. 36 sq. cm.

Daily Review 15
1. (C) 90° 2. 64 sq. m. 3. $5.20 4. $12
5. (D) 150,000

Daily Review 16
1. 5 2. (B) cups 3. (C) cone 4. 12:35 5. 12

Daily Review 17
1. $1.50 or 150¢ 2. $2.05 or 205¢ 3. cloudy
4. 15 5. (C) 20

Daily Review 18
1. $400 2. food and housing 3. (B) 2000
4. (D) 18,000 5. (C) 200

Pre- & Post Tests, Daily Reviews — Answer Keys

Daily Review 19
1. $1\tfrac{5}{8}$ 2. $\tfrac{12}{13}$ 3. > 4. $1\tfrac{3}{5}$ 5. $\tfrac{11}{8} = 1\tfrac{3}{8}$

Daily Review 20
1. $3\tfrac{4}{5}$ 2. $\tfrac{3}{4}$ 3. $1\tfrac{7}{8}$ 4. 6 5. $\tfrac{2}{15}$

Daily Review 21
1. 1.2 2. 0.018 3. thousandths 4. 0.89, 1.18, 1.2, 1.25 5. 0.31

Daily Review 22
1. 0.775 2. 54.3 3. 85% 4. 21% 5. $1.11

Daily Review 23
1. (one) millions 2. (C) 3. 5000 4. 1, 2, 3, 6, 9, 18 5. 40

Daily Review 24
1. 13,532 2. 5,066 3. 26,391 4. 612 R2 5. 126

Daily Review 25
1. $\tfrac{5}{8}$ 2. $\tfrac{3}{4}$ 3. $\tfrac{1}{6}, \tfrac{1}{3}, \tfrac{1}{2}, \tfrac{5}{6}, \tfrac{3}{3}$ 4. $2\tfrac{1}{4}$ 5. $\tfrac{2}{3}$

Daily Review 26
1. $6\tfrac{3}{4}$ 2. $\tfrac{14}{15}$ 3. $2\tfrac{5}{8}$ 4. $\tfrac{4}{15}$ 5. $\tfrac{5}{24}$

Daily Review 27
1. 0.7 2. five and forty-three hundredths 3. 3 4. 0.59 5. >

Daily Review 28
1. 3.567 2. 0.2958 3. 4.6 4. 100% 5. $\tfrac{81}{100}$

Daily Review 29
1. (D) line segment 2. (D) 3. (B) 4. (D) 5. (B)

Daily Review 30
1. (B) $1\tfrac{1}{4}$ 2. (B) 45° 3. 18 sq. cm. 4. 60 cu. in. 5. 4:45

Daily Review 31
1. (C) pounds 2. 4 3. $1.79 4. 20 5. 6¢

Daily Review 32
1. 14 2. 80 3. 20 4. (B) 8000 5. (D) 35,000

Skill Builders — Answer Keys

1-1
1. 2, 8, 1 **2.** 7, 6, 3 **3.** 0, 2, 9 **4.** 6, 5, 4 **5.** ten billions 80,000,000,000 **6.** one million 8,000,000 **7.** one billions 1,000,000,000 **8.** hundred billions 900,000,000,000 **9.** 6 **10.** 0 **11.** 2 **12.** 8 **13.** 1 **14.** 0 **15.** 7 **16.** 5 **17.** 3 **18.** 4 **19.** 9 **20.** 2

2-1
1. < **2.** > **3.** < **4.** < **5.** > **6.** < **7.** > **8.** > **9.** < **10.** < **11.** > **12.** > **13.** 13,849 13,872 13,881 **14.** 1,254,843 1,254,879 1,254,986 **15.** 19,125 19,365 20,006 **16.** 870,622 870,643 870,980 **17.** 2,643,569 2,643,596 2,727,128 **18.** 55,783,296 55,783,297 55,783,299

3-1
1. 7000 8000 7500 less 7000 **2.** 4000 5000 4500 greater 5000 **3.** 1000 2000 1500 1000 **4.** 9000 10,000 9500 9000 **5.** 3000 4000 3500 3000 **6.** 9000 10,000 9500 10,000 **7.** 0 1000 500 1000 **8.** 4000 5000 4500 5000 **9.** 0 1000 500 0 **10.** 12,000 13,000 12,500 12,000 **11.** 39,000 40,000 39,500 40,000 **12.** 115,000 116,000 115,500 115,000

3-2
HE BUTTERED UP THE BOSS

4-1
Shaded numbers: 2, 3, 5, 7, 11, 13, 17, 19, 23, 29, 31, 37

5-1
1. 56 **2.** 107,212 **3.** 6 **4.** 6 **5.** 8 **6.** 10 **7.** 42, 24, 168, 168 **8.** 946 **9.** 5 **10.** 6 **11.** 31 **12.** 12 **13.** 18, 54, 108, 108 **14.** 5 × 4, 4 × 9, 180, 180
PUZZLE: associative property lets you multiply the middle number by 0 first; the answer is 0.

5-2
1. 6, 8, 24, 24 **2.** 20, 10, 100, 100 **3.** 4 **4.** 7 **5.** 10 **6.** 2 **7.** 2 **8.** 1 **9.** 4 **10.** 6 **11.** Ex: (5 × 6) × 3 = 5 × (6 × 3) **12.** Ex: (5 × 6) × 3 = 5 × (6 × 3) **13.** a, c **13.** 5 × 6 × 7 = 210

6-1
1. 37,653 **2.** 7806 **3.** 9052 **4.** 291,563 **5.** 70,610 **6.** 94,357 **7.** 322,974 **8.** 16,130 **9.** 66,611 **10.** 70,006 **11.** 56,452 **12.** 61,624

7-1
1. 263,115 **2.** 483,348 **3.** 572,337 **4.** 792,294 **5.** 670,453 **6.** 423,180 **7.** 392,179 **8.** 221,271

8-1
1. 36,800 **2.** 20,288 **3.** 6188 **4.** 9246 **5.** 17,112 **6.** 13,132 **7.** 66,810 **8.** 19,845 **9.** 47,383 **10.** 39,812 **11.** 66,068 **12.** 20,875 **13.** 4029 **14.** 17,184 **15.** 7840 **16.** 7839

8-2
MORE THAN THE STUDENTS

9-1
1. 2102 **2.** 749 **3.** 941 R6 **4.** 925 **5.** 763 R5 **6.** 876 R3 **7.** 751 R1 **8.** 143 **9.** 23,292 **10.** 24,655 R2 **11.** 7719 R1 **12.** 16,400 R5 **13.** 9122 **14.** 3949 R2 **15.** 2256 **16.** 3442 **17.** 44,873 R2 **18.** 57,938 **19.** 64,353 **20.** 488,216

10-1
1. 20)320 Answer: 16 **2.** 30)360 Answer: 12

3. 10)170 Answer: 17 **4.** 20)2640 Answer: 132

5. 14 **6.** 45 **7.** 13 **8.** 12 **9.** 12 **10.** 12 **11.** 15 **12.** 14 **13.** 240 **14.** 35 **15.** 46 **16.** 150 **17.** 48 **18.** 26 **19.** 24 **20.** 36

10-2
1. 3 **2.** 4 R11 **3.** 7 R3 **4.** 42 R17 **5.** 12 **6.** 18 **7.** 13 **8.** 6 **9.** 11 R1 **10.** 27 **11.** 10 R65 **12.** 27 R5 **13.** 4 R32 **14.** 14 R1 **15.** 44 R2 **16.** 12 R8 **17.** 13 **18.** 21 R10 **19.** 3 R12 **20.** 14 Shaded figure is a pair of sunglasses.

Skill Builders — Answer Keys

10-3
1. 24 2. 33 R14 3. 4 R11 4. 29 5. 4 R42
6. 21 R10 7. 4 R29 8. 6 9. 51 R12 10. 52
11. 43 R6 12. 83 13. 74 14. 38 R52 15. 11 R41 16. 41

11-1
1. 8, 3, $3/8$ 2. 10, 3, $3/10$ 3. $4/5$ 4. $5/6$ 5. $11/12$ 6. $0/4$
7. 4, 3, $3/4$ 8. 8, 5, $5/8$ 9. $1/6$ 10. $2/3$ 11. $4/4$ 12. $2/5$
13. $7/10$ 14. $7/12$

11-2
1. $1/4, 5/4$ 2. $5/6, 5/5$ 3. $4/2$ 4. $7/12, 3/10$ 5. $3/3, 2/2$
6. $5/5$ 7. $5/4$ 8. $4/3, 10/5$ 9. 2 10. $2\,1/4$

12-1
1. $3/4$ 2. $1/2$ 3. $4/6$ 4. $2/5$ 5. $1/2$ 6. $3/4$ 7. $1/1$ 8. $5/6$
9. $1/3$ 10. $1/3$ 11. $2/3$ 12. $2/3$ 13. $2/3$ 14. $1/7$ 15. $3/4$
16. $1/2$ 17. $2/5$ 18. $3/8$ 19. $1/3$ 20. $1/3$

12-2
SIMPLEST FRACTION BINGO

13-1
1. $0/4, 1/4, 2/4, 3/4, 4/4$ 2. $0/8, 1/8, 2/8, 3/8, 4/8, 7/8$ 3. $1/6, 2/6, 1/2, 5/6, 6/6$ 4. $1/10, 3/10, 1/2, 7/10, 10/10$ 5. $0/12, 4/12, 1/2, 7/12, 12/12$ 6. $1/5, 2/5, 1/2, 3/5, 5/5$ 7. $1/8, 1/4, 3/8, 1/2, 5/8$ 8. $1/8, 1/2, 5/8, 3/4, 7/8$ 9. $1/5, 1/4, 1/3, 1/2, 1/1$ 10. $1/12, 1/10, 1/8, 1/5, 1/2$ 11. $0/6, 1/6, 1/3, 1/2, 2/3$ 12. $1/12, 1/6, 1/3, 1/2, 2/3$ 13. $1/12, 1/6, 1/4, 1/3, 5/12$ 14. $1/2, 7/12, 2/3, 3/4, 11/12$ 15. $1/10, 3/10, 2/5, 1/2, 3/5$ 16. $1/4, 3/8, 1/2, 3/4, 7/8$

14-1
WITH A SORE THROAT?
A CENTIPEDE WITH SORE FEET

15-1
1. $2/3$ 2. $2/5$ 3. $1/2$ 4. $2/5$ 5. $1/6$ 6. $1/3$ 7. $1/6$ 8. $1/3$
9. $4/15$ 10. $1/10$ 11. $1/3$ 12. $1/4$ 13. $1/2$ 14. $1/4$
15. $1/2$ 16. $1/4$ in.

16-1
1. $3\,1/2$ 2. $3\,1/2$ 3. $4\,1/4$ 4. $2\,1/3$ 5. $2\,1/2$ 6. $3\,2/3$
7. $3\,3/4$ 8. $3\,3/4$ 9. $2\,3/4$ 10. $3\,2/3$ 11. $4\,2/3$ 12. $4\,2/5$
13. $3\,5/12$ 14. $7\,1/4$ 15. $5\,1/2$ 16. $3\,1/3$ 17. $5\,1/2$ c.
18. $4\,4/5$ mi.

16-2
1. $1\,3/8$ 2. $2\,3/4$ 3. $1\,2/3$ 4. $1\,3/5$ 5. $1\,7/10$ 6. $2\,5/6$
7. $1\,7/12$ 8. $2\,3/8$ 9. $4\,1/3$ 10. 6 11. $6/7$ 12. $11\,1/10$
13. $2\,1/2$ 14. $1\,1/6$ 15. $2\,1/2$ 16. $2\,3/4$ 17. $1\,5/8$
18. $1\,1/3$ 19. $2\,5/6$ 20. $3\,1/2$ ft.

17-1
TO A NIGHT CLUB

17-2
1. $5/12$ 2. $14/15$ 3. $13/24$ 4. $17/18$ 5. $17/35$ 6. $23/24$
7. $23/30$ 8. $5/24$ 9. $7/12$ 10. $11/24$ 11. $7/22$ 12. $2/15$
13. $1/9$ 14. $7/36$ 15. $7/12$ 16. $7/20$

18-1
1. $3\,3/4$ 2. $5\,5/6$ 3. $3\,7/8$ 4. $2\,11/12$ 5. $4\,7/8$ 6. $3\,5/12$
7. $3\,7/10$ 8. $3\,7/10$ 9. $6\,8/15$ 10. $5\,17/24$ 11. $5\,7/24$
12. $7\,3/5$ 13. $2\,9/10$ mi. 14. $3\,5/6$ yd. 15. 4 lb.
16. $1\,11/16$ in.

19-1
1. $1/8$ 2. $1/2 \times 1/3 = 1/6$ 3. $1/2 \times 1/5 = 1/10$
4. $1/2 \times 1/8 = 1/16$ 5. $1/12$ 6. $1/3 \times 1/2 = 1/6$
7. $1/3 \times 1/5 = 1/15$ 8. $1/3 \times 1/6 = 1/18$ 9. $1/4 \times 1/2 = 1/8$
10. $1/4 \times 1/3 = 1/12$

19-2
1. $1/2$ 2. 8 3. 10 4. $2/3$ 5. 4 6. 3 7. 3 8. 3
9. 3 10. 5 11. 8 12. 2 13. 12 14. 7 15. 75
16. 6 17. 1 18. 2 19. 4 20. 5 21. 1 c.
22. 60¢ left

20-1
1. $4/3$ 2. $5/4$ 3. $1/5$ 4. $1/7$ 5. $3/1$ 6. $8/5$ 7. $1/8$ 8. $1/4$
9. 6 10. 12 11. $5/16$ 12. $1/24$ 13. $8/9$ 14. $3/5$
15. $15/16$ 16. $3/4$ 17. $5/14$ 18. $9/10$ 19. $32/45$ 20. $4/15$
21. 18 22. 20

21-1
1. $3/100$, 0.03 2. $7/100$, 0.07 3. $7/100$, 0.07 4. $10/100$ or $1/10$, 0.10 or 0.1 5. $100/100$ or $1/1$, 1.0 6. 3 of 100 small squares shaded 7. 61 of 100 small squares shaded 8. 19 of 100 small squares shaded

Skill Builders — Answer Keys

22-1
1. $2^{7}/_{10}$, 2.7, two and seven tenths 2. $1^{23}/_{100}$, 1.23, one and twenty-three hundredths 3. 1.4
4. 5.21 5. 2.9 6. 3.75 7. thirteen hundredths
8. five tenths 9. one and seven tenths 10. six and twelve hundredths 11. twenty-one and seven tenths 12. four and five hundredths
13. thirty hundredths or three tenths 14. nine and four tenths 15. two and sixty one hundredths

23-1
1. 6 2. 2 3. 1 4. 5 5. 4 6. 9 7. 3 8. 0 9. 5
10. 6 11. tenths 12. hundredths 13. ones
14. tens 15. hundredths 16. hundreds 17. 0.27
18. 1.76

23-2
1-6. Be sure that only one digit goes in each box and that the decimal point separates the whole number and the decimal part. 7. thousandths
8. hundredths 9. hundreds 10. tenths
11. ten thousandths 12. tens 13. thousands
14. thousandths 15. ten thousandths
16. hundredths

24-1
1. 0.6 < 0.65 2. 0.40 = 0.4 3. > 4. = 5. <
6. = 7. > 8. < 9. < 10. > 11. > 12. < 13. <
14. > 15. 0.45, 0.5, 0.52 16. 0.09, 0.2, 0.4

25-1
ON THE FACE OF A WATCH!

26-1
TAKE AWAY THE "S"

27-1
1. 0.6 2. 1.2 3. 1.5 4. 1.8 5. 0.48 6. 0.72
7. 2.52 8. 10.16 9. $1.05 10. $15.60
11. $28.80 12. $51.10 13. $1.92 14. $14.22
15. $88.50 16. $29.75 17. $18.00 18. $76.00

27-2
1. $^{3}/_{10} \times {}^{2}/_{10} = {}^{6}/_{100}$, 0.06 2. $^{5}/_{10} \times {}^{3}/_{10} = {}^{15}/_{100}$, 0.15 3. $^{7}/_{10} \times {}^{31}/_{100} = {}^{217}/_{1000}$, 0.217
4. $^{32}/_{100} \times {}^{4}/_{100} = {}^{128}/_{10000}$, 0.0128 5. 1.86 6. 0.192
7. 0.0192 8. 0.60 9. 0.185 10. 0.3 11. 6.66
12. 0.0476 13. 0.3 14. 0.1344 15. 4.8 16. 4.06

28-1
1. 0.27 2. 0.29 3. 0.13 4. 0.12 5. 0.08 6. 0.06
7. 0.05 8. 1.44 9. $2.35 10. $1.35 11. $0.41
12. $1.12 13. $0.45 14. $3.25 15. 12¢ 16. $0.40

28-2
1. 14 2. 32 3. 12 4. 19 5. 9 6. 2.3 7. 4.3
8. 14 9. 3 10. 0.5 11. 1.35 12. 33.5 13. 32
14. 50 15. 8 16. 64

29-1
1. 3¢, 3% 2. 7¢, 7% 3. 13¢, 13% 4. 29¢, 29%
5. 17% 6. 64% 7. 85% 8. 21 small squares shaded 9. 75 small squares shaded 10. 100 small squares shaded 11. No, the numbers add to 110% 12. No, the numbers add to 105%
13. Yes, the numbers add to 100%

30-1
1. $^{27}/_{100}$ 2. $^{91}/_{100}$ 3. $^{13}/_{100}$ 4. $^{7}/_{100}$ 5. $^{1}/_{100}$
6. $^{16}/_{25}$ 7. $^{1}/_{1}$ 8. $^{1}/_{5}$ 9. $^{1}/_{2}$ 10. $^{1}/_{4}$ 11. $^{3}/_{4}$ 12. $^{9}/_{10}$
13. 24% 14. 50% 15. 8% 16. 10% 17. 85% 18. 1%
19. 17% 20. 69% 21. 100% 22. 50% 23. 25%
24. 75% 25. A = 50% B = 25% C = 25%
26. A = 75% B = 12.5% C = 12.5%

31-1
Figures 2 and 4 are crossed out. 6. ∠ABC, ∠CBA, ∠B 7. ∠XYZ, ∠ZYX, ∠Y 8. ∠LMN, ∠NML, ∠M 9-11. Angle size will vary; the vertex must be the middle letter.
PUZZLE: 15 angles

32-1
1-2. Answers will vary. 3–5. Lines at 90° angles will be drawn 6. perpendicular, parallel
7. parallel, perpendicular 8. ST and SQ, SQ and QR, QR and RT, RT and TS

33-1
1. acute, right, or obtuse 2. right 3. obtuse
4. right 5. acute 6. right 7. obtuse 8. right
9. acute 10. obtuse 11. obtuse 12. acute
13. right

Skill Builders — Answer Keys

34-1
1. quadrilateral (or trapezoid) 2. hexagon
3. triangle (or right triangle) 4. octagon
5. pentagon 6. quadrilateral (or parallelogram)
7. decagon 8. triangle (or equilateral triangle)
9. square 10. 5 of each 11. 8 12. 3

35-1
1. circumference 2. radius 3. diameter
4. circumference 5. radii (plural of radius)
6. 7, 14 7. 6, 12 8. 4.5, 9.0 9. 40 feet
10. They are equal.

36-1
2. All measurements are approximate and rounded to the nearest inch. line AB = 3 in. line CD = 3 in. line EF = 5 in. line GH = 1 in. line KL = 4 in. 3. All measurements are approximate and rounded to the nearest ½ inch. line MN = 1½ in. line PQ = 1 in. RS = 2½ in. line TU = ½ in. line VW = 3½ in. line XY = 4 in.

36-2
Ruler is marked: 0, ⅛, ¼, ⅜, ½, ⅝, ¾, ⅞, 1, 1⅛, etc. line CD = 4⅞ line EF = ⅞ in. line PQ = 2½ in. line ST = 3⅛ in. line VW = 3⅝ in. 0/8, ⅛, 2/8, ⅜, 4/8, ⅝, 6/8, ⅞, 8/8, 9/8, ... 24/8 0, ⅛, ¼, ⅜, ½, ⅝, ¾, ⅞, 1, 1⅛, 1¼, 1⅜, 1½, 1⅝, 1¾, 1⅞, 2

37-1
Check angles with a protractor

38-1
1. 18 cm 2. 16 cm 3. 12 cm 4. 22 cm 5. 20 cm 6. 29 cm

38-2
1. 18 2. 9 3. 12 4. 24 5. 16 6. 14 7. 14
8. 11 9. 4 10. 4½ 11. 6

39-1
1. 20 2. 24 3. 100 4. 1000

39-2
1. 6 2. rectangles (or 4 rectangles and 2 squares) 3. 12 4. 8 5. 4 6. 4 equilateral triangles 7. 6 8. 4

40-1
1. 120 2. 3 3. 36 4. 35 5. 48 6. 2⅙ 7. 90
8. 132 9. 160 10. 310 11. ½ 12. ¾ 13. 1½
14. 3⅓ 15. 5 hr. 45 min. 16. 3 hr. 50 min.
17. 9 hr. 18. 2 hr. 20 min. 19. 2 hr. 27 min.
20. 1 hr. 45 min. PUZZLE: Answers will vary.

41-1
1. grams 2. kilograms 3. grams
4. kilograms 5. kilograms 6. grams 7. 2000
8. 3 9. 4000 10. 3500 11. 2.5 12. 1.3 13. 0.5
14. 1250 15. 0.2 16. 2300 17. 1.75 18. 3620
PUZZLE: "Kilo" means thousand. There are 1000 liters in a kiloliter.

42-1
1. 4 2. 8 3. 1 4. 8 5. 2 6. 8 7. 4 8. 6 9. 16
10. 1 11. 6 12. 3 13. 1½ 14. 6 15. ½ 16. 8 gal. 17. 24 c. 18. $6.80 19. 16 c.

43-1
1. 28¢ 2. $3.10 3. $7.65 4. $7.38 is the correct change from a $10 bill. 5. $3.56 6. $3.85
7. $9.15 8. $6.21 9. $1.63 10. 27¢

44-1
1. 23, 27, 31 2. 47, 42, 37 3. 64, 57, 50, 43
4. 24, 30, 36 5. 13, 17 (pattern of primes)
6. 8, 13 (add 2 preceding digits)
Puzzle piece: 6, 9, 7, 10, 8, 15, 13, 18, 16, 21

45-1
STAGE COACH

45-2
All strategies could be used in each problem. Estimates may vary. 1. mult. 216 2. subt. 356
3. mult. $935 4. div. 34 5. add 123 6. div. 16

45-3
1. 10 2. ⅔ 3. ⅗ 4. ⅛ in. 5. ⅜ 6. 5½ yd.
7. 1/4 8. 6¼ yd.

Assessment 23

Skill Builders
Answer Keys

45-4
1. 25 card box is 15¢ per card cheaper.
2. 6 bars is 5¢ per bar cheaper. 3. The 10 ounce bag is $0.03 per ounce cheaper. 4. The 6 pack is 10¢ per bar cheaper. 5. The 100 napkins are $0.005 cheaper per napkin.
6. The 40 oz. jar is $0.01 per ounce cheaper.
7. The ½ dozen is $0.05 per doughnut cheaper.
8. The dozen oranges is $0.05 per orange cheaper.

45-5
1. $n = 6 \times 13$, $n = 78$ miles 2. $n = 584 \div 12$, $n = \$48.67$ 3. $n = 26\text{-}18$, $n = 8$
4. $n = 1{,}708{,}652 + 1{,}700{,}600$, $n = 3{,}409{,}252$ visitors

46-1
1. 10 cm 2. 10 cm 3. 4 cm 4. 8 cm 5. 9 cm
6. 16 cm

46-2
ON THE AVERAGE HE'S COMFORTABLE

47-1
1. 235 2. 645 3. 455 4. 341 5. any combination that equals over 400 calories
6. any combination that equals less than 200 calories 7. 34,560 8. 27 min.

47-2
1. Less than likely 2. 6/36 or 1/6 3. 30/36 or 5/6 4. Give the player 5 points instead of 3 points.

48-1
1. 10 2. 30 3. 20 4. weeks 4 and 8 5. 5 wpm is a good estimate 6. 50

49-1
1. (D) 2. (D) 3. 60 100 70 110 70 110
4. 50 90 120 160 100 140 5. 600 800 800 1000 900 1100 6. 800 1000 700 900 600 800
7. 6000 10,000 6000 10,000 5000 9000 8. 7000 9000 6000 8000 10,000 12,000 Estimates may vary. Accept reasonable estimates.

49-2
1. (C) 2. (A) 3. (D) 4. 40 5. 50 6. 40 7. 200
8. 400 9. 500 10. 4000 11. 3000 12. 5000

50-1
1. $7 \times 50 = 350$ 2. $8 \times 50 = 400$
3. $9 \times 80 = 720$ 4. $90 \times 6 = 540$
5. $4 \times 100 = 400$ 6. $70 \times 3 = 210$
7. $8 \times 50 = 400$ 8. $20 \times 6 = 120$
9. $300 \times 2 = 600$ 10. $3 \times 300 = 900$
11. $7 \times 80 = 560$ 12. $6 \times 20 = 120$
13. $4 \times 20 = 80$ 14. $200 \times 6 = 1200$
15. $80 \times 5 = 400$ 16. $3 \times 90 = 270$
17. $9 \times 400 = 3600$ 18. $7 \times 800 = 5600$
19. $60 \times 6 = 360$ 20. $9 \times 20 = 180$

50-2
1. $20 \times 50 = 1000$ 2. $50 \times 10 = 500$
3. $30 \times 50 = 1500$ 4. $30 \times 80 = 2400$
5. $60 \times 80 = 4800$ 6. $60 \times 30 = 1800$
7. $60 \times 50 = 3000$ 8. $10 \times 80 = 800$
9. $70 \times 80 = 5600$ 10. $60 \times 40 = 2400$
11. $70 \times 300 = 21{,}000$ 12. $50 \times 400 = 20{,}000$
13. $90 \times 300 = 27{,}000$ 14. $60 \times 500 = 30{,}000$
15. $40 \times 400 = 16{,}000$ 16. $30 \times 600 = 18{,}000$
17. $20 \times 300 = 6{,}000$ 18. $90 \times 400 = 36{,}000$
19. $50 \times 200 = 10{,}000$ 20. $70 \times 600 = 42{,}000$

21.
300	1200
400	1600
700	2800

22.
1800	9000
4800	24000
3600	18000

23.
14000	16000
28000	32000
28000	32000

Glossary of Math Terms — Answer Keys

acute angle... any angle that measures less than 90°

addition... the operation that combines two or more numbers

area... the number of square units needed to cover the surface of a figure

associative (grouping) property... changing the grouping of addends (or factors) does not change the sum (or product)

average... the quotient obtained by dividing a sum by the number of addends

bar graph... a graph of data that uses lengths of bars to show the information

capacity... the maximum amount that can be contained by an object

center... the point that is equidistant from all points on a circle

circle... a set of points in a plane, all of which are the same distance from a given point, called the center

circumference... the distance around a circle, about 3.14 times the diameter

closed figure... a figure that begins and ends at the same point

common factor... a number that is a factor of two or more whole numbers

commutative (order) property... changing the order of the addends (or factors) does not change the sum (or product)

compass... an instrument used to draw circles

congruent... figures that have the same size and shape

coordinate points... an ordered pair of numbers that identify a point on a coordinate plane

data... numerical information

decimal fraction... a fraction expressed in decimal notation whose denominator is a power of 10

degrees (°)... a unit used to measure angles

denominator... the number written below the fraction bar that tells how many equal parts are in the whole

diagonal... lines having a slanted direction

diameter... a line segment that passes through the center of a circle and has both endpoints on the circle

differences... having qualities that are not alike

discount... a reduction in the regular, or list, price of an item

dividend... the number to be divided

division... an operation that divides a set, region, or number into equal parts

divisor... the number by which another number is to be divided

dozen... a group of 12

edge... the line segment where two faces of a solid figure meet

equivalence... having characteristics of equality

equivalent fractions... two or more fractions that represent the same number

estimate... to find an approximate solution mentally by using rounded numbers

face... a plane figure that serves as one side of a solid figure

factor... a number that divides evenly into another

flip... a transformation creating a mirror image of a figure on the opposite side of a line; also called a *reflection*

Glossary of Math Terms — Answer Keys

fraction... a number used to name a part of a group or a whole

greatest common factor... the greatest number that is a factor of two or more numbers

horizontal... lines running across from left to right

improper fraction... a fraction with its numerator equal to or greater than its denominator

intersecting... lines that cross at a point

leading digit... the first digit in a number, usually the number rounded to for estimation

least common denominator... the least number that is a common denominator of two or more fractions, also called the *least common multiple*

line graph... a pictorial representation of data that uses connected line segments to show changes over time

line segment... a part of a line that has two endpoints

lowest terms... when the greatest common factor of the numerator and the denominator of a fraction is 1

mean... an average; the sum of all the data points, divided by the number of data points

mixed number... a number written as a whole number and a fraction

multiplication... the operation of repeated addition

numerator... the number written above the fraction bar that tells how many equal parts of the whole are being counted

obtuse angle... any angle whose measure is greater than 90° and less than 180°

open figure... a figure that does not begin and end at the same point

ordered pair... a pair of numbers used to locate a point in a coordinate plane

parallel... lines in a plane that do not intersect

parallelogram... a quadrilateral with two pairs of parallel and congruent sides

percent... a special ratio that compares a number to 100 using the symbol %

perimeter... the distance around a polygon, found by adding the lengths of all the sides

perpendicular... intersecting lines that meet at square angles

pictograph... pictorial representation of data that uses a single symbol to represent multiples of a quantity

place value names to hundred billions place... ones, tens, hundreds, thousands, ten thousands, hundred thousands, millions, ten millions, hundred millions, billions, ten billions, hundred billions

place value decimal names to ten-thousandths place... tenths, hundredths, thousandths, ten-thousandths

price per unit... the cost of a single unit of an item

prime number... a whole number greater than 1 that has only two factors, itself and 1

probability... the chance of an event occurring; equal to the number of favorable outcomes divided by the number of possible outcomes; a measure of chance; probability ranges from 0 (impossible event) to 1 (certain event)

product... the result of multiplication

proper fraction... a fraction whose numerator is less than its denominator

property... a rule that is always true

protractor... a tool used to measure angles

Glossary of Math Terms — Answer Keys

quadrilateral... a four-sided polygon

quotient... the number that results from dividing

radius... a line segment from the center of a circle to any point on the circumference

range... the difference between the greatest and the least numbers in a data set

reciprocal... two numbers that have a product of 1, e.g., $\frac{1}{3}$ and $\frac{3}{1}$

regular decagon... a 10-sided polygon with all sides and all angles congruent

regular hexagon... a 6-sided polygon with all sides and all angles congruent

regular octagon... a 8-sided polygon with all sides and all angles congruent

regular pentagon... a 5-sided polygon with all sides and all angles congruent

regular polygon... a polygon with all sides and all angles congruent

right angle... an angle, a square corner, that measures exactly 90°

rounding... to approximate a number by replacing it with a number expressed to the nearer ten, hundred, thousand, and so on

set... a well-defined group of objects

similarities... having characteristics in common or alike

simplify... to change to lowest terms

slide... a transformation that slides a figure a given distance in a given direction, also called a *translation*

square... a rectangle with all sides congruent

straight angle... an angle that measures 180°

subtraction... an arithmetic operation that takes away a given amount or compares two numbers

sum... the result of addition

triangle... a 3-sided polygon

turn... a transformation in which a figure is turned a given angle and direction around a point, also called a *rotation*

unlike fractions... fractions that have different denominators

Venn diagram... a drawing that shows relationships among sets of objects

vertex... the point at which two line segments, lines or rays meet to form an angle

vertical... lines running straight up and down

volume... the number of cubic units needed to fill the space inside a figure

weight... the heaviness of an object

Glossary of Math Terms — Answer Keys

LENGTH

Metric

1 kilometer = 1000 meters
1 meter = 100 centimeters
1 centimeter = 10 millimeters

Customary

1 mile = 1760 yards
1 mile = 5280 feet
1 yard = 3 feet
1 foot = 12 inches

CAPACITY & VOLUME

Metric

1 liter = 1000 milliliters

Customary

1 gallon = 4 quarts
1 gallon = 128 ounces
1 quart = 2 pints
1 pint = 2 cups
1 cup = 8 ounces

MASS AND WEIGHT

Metric

1 kilogram = 1000 grams
1 gram = 1000 milligrams

Customary

1 ton = 2000 pounds
1 pound = 16 ounces

TIME

1 millenium = 1000 years
1 decade = 10 years
1 year = 12 months
1 week = 7 days
1 hour = 60 minutes

1 century = 100 years
1 year = 365 days
1 year = 52 weeks
1 day = 24 hours
1 minute = 60 seconds

Journal Prompt Instructions and Scoring Guide

Journal Prompts are integrated into the Teacher Guide lessons to assess knowledge at specific points in the curriculum. (See pp. 3, 4, 7, 10, 14, 18, 21, 23, 26, 31, 32, 38, 39, 45, 49, 52, 56, 58 and 63.) The assessments ask a student to show their understanding of a concept by using it to solve a problem and show the solution in different ways, such as with numbers, diagrams and words. The blank forms on the following pages may be used as the cover and insides of students' writing journals.

Teachers may use the following guide to assign scores. An example of a journal prompt and two samples for each score follows.

Tips for Scoring:

1. A score of 1 generally means that the student demonstrates very little understanding of the task.
2. A score of 2 generally shows understanding, but is incomplete in some way.
3. A score of 3 will not usually be perfect. Students need only to complete the task correctly and demonstrate that they understand.
4. In assigning the score, teachers should consider:
   - Use of pictures and diagrams
   - Whether work is shown or labeled
   - Correct use of mathematical words
   - Clarity of the explanation
5. Grammar and punctuation should **NOT** be considered in determining the score.

 Journal Prompt: What is a perimeter? When can it be used in the real world?

Examples of 1-Point Scores

The student shows limited understanding and is unable to communicate mathematically.

> A perimeter is like you have to mearsure the size then you can use it to measure a door.

This student recognizes that perimeter is a type of measurement, but does not distinguish it from other measurements such as area or height. This score is a "high" 1.

> What's a perimeter it is a a that to put number
>
> On a door

This student only recognizes that perimeter uses numbers. The real-world example is apparently a door. The response is disjointed and unclear. This score is a "low" 1.

Examples of 2-Point Scores
The student partially completes the task and/or communication is limited.

① What is a perimeter

② A perimeter is a shapes that have number on the side you add

When can it be in the real world

I can used it when I making a fnece with by measure

This student writes that perimeter has to do with adding the numbers around a shape. The real-world example is measuring a fence, but it does not contain any further explanation. This score is a 2.

① What is a perimeter it to add the number at the side together.

When can it be in the real World? I could use it to measure when I am racing with my friends.

This student correctly notes that finding the perimeter means to add the numbers on the sides together. The real-world example is unclear. This score is a 2.

Examples of 3-Point Scores
The student completes the task correctly and communicates effectively.

What is perimeter? A perimeter is a shape and you add up all the sides

4 6 4+6+3=13
3
add all sides

When can it be used in the real world? It could be measure on a gate by useing it with a ruler and do the sides.

This student writes that perimeter is adding the sides of a shape together. The picture clearly demonstrates knowledge of the concept. The real-world example is making a gate by using a ruler to measure the sides. This score is a 3.

perimeter is length around a shape. you count the sides on the perimeter and you add it up. when can it be used in the real world?

we can use it when people is putting up a gate. You have to measure the gate to see if the gate can fit in the spot.

This student has the best definition of a perimeter – "length around a shape" – and states that it means to add the sides of a shape. The example of a gate notes that measuring allows a person to see if the gate will fit. This score is a 3.

My Math Journal

"How do I know what I think until I see what I say?"

Name _____
School _____

Journal Prompt

Question: _____

Answer: _____

Remember to show your work and draw pictures to help explain your answer.

Journal Prompt

Question:

Answer:

Remember to show your work and draw pictures to help explain your answer.

Journal Prompt

Question:

Answer:

Remember to show your work and draw pictures to help explain your answer.

Teacher Guide

The Teacher Guide section contains lesson plans for each page in the student book.

Achievement improves significantly when teachers are well-organized. Use the pacing calendars found on pages *ix* to *xii* of the Foreword for a 20-day session.

To prepare calendars for a different number of days, make copies of the blank calendar on page *xiii*. Decide about how many pages you want to cover each day. Plan to complete all pages about two days before the end of the extended school session, so there is adequate time to complete the Post-Test. You will also need to cover the 32 Daily Reviews. Use as many Skill Builders as needed to provide flexibility for your extended school time.

Objective: To discover the pattern in the Hindu-Arabic number system. To identify the place value of numbers up to 12 digits in length. **DVD**

Materials: Base ten blocks, Place Value Mat (Masters 10 and 11), posterboard for making Place Value classroom chart, tape, Centimeter Graph Paper (Master 2)

Vocabulary: place value names to hundred billions place

Introductory Activities

Discovering the Pattern

Give each group of 2 to 4 students a set of base ten blocks: **Today we will explore with a set of base ten blocks.** Have one person in each group record the similarities and differences.

Similarities	Differences
made of wood	sizes
natural color	shapes
points and corners	volumes
solids	weight
made of 1 cm cubes	
10 of 1 block = 1 of the next larger	

Discuss the similarities and differences found. Emphasize that 10 of any one block has the same value as the next larger block.

Name the blocks in order of size as ones, tens, hundreds and thousands, placing each block in the appropriate place on a Place Value Mat (copies of Master 10 and 11 taped together). **If we put 10 one thousands cubes together, what is the name of the next place value?** (ten thousand)

To make a model of ten thousand, use Centimeter Graph Paper (Master 2) to build a model of 10 one thousand cubes and tape them together. **If we put 10 ten thousands together, what is the name of the next place value? Count by ten thousands with me.** (10,000, 20,000, ... 100,000) Draw a chart of the first 6 place values.

If we put 10 of the very large blocks together, what is the name of the next place value? Count by hundred thousands to name the new place value. (100,000, 200,000, 300,000, ... 1,000,000)

If we put 10 one millions together, what is the name of the next place value? Count with me. (1 million, 2 million, ... 10 million)

If we put 10 ten millions together, what is the name of the next place value? Count with me to name the next place value. (10 million, 20 million, ... 90 million, 100 million)

Ten hundred million is the same as 1 billion, 10 ten billions is the same as 1 hundred billion. Write each of the 3 new place values on the classroom chart.

Use a colored pen to outline the words "one," "ten" and "hundred" in each group of 3 place values. **The place value names can be thought of as family names for numbers. In number families, we have different first names; the first names of the place value families are always "one," "ten" and "hundred."** Use a different color pen to outline the words "thousands," "millions" and "billions." **Every group of 3 place values has the same last name. The last names are the thousands, millions, and billions.**

About This Page

Refer to the chart at the top of the page. Ask a student to read the number in the place value chart.

Follow Up Activities

Place Value Bingo (Master 1)

Skill Builders 1-1

A Place Value Chart to Billions
The national debt in the United States in 1991 is shown in the place value chart.

Billions			Millions			Thousands			Units		
Hundred Billions	Ten Billions	Billions	Hundred Millions	Ten Millions	Millions	Hundred Thousands	Ten Thousands	Thousands	Hundreds	Tens	Ones
		3	6	8	3	0	0	0	0	0	0

Look at the number in the place value chart above.
What digit is in each place?

1. billions __3__
2. hundred millions __6__
3. ten millions __8__

Write the place name and value of each underlined digit.

		Place Name	Value
4.	6,473,758,216	one billions	6,000,000,000
5.	26,000,000,000	ten billions	20,000,000,000
6.	761,000,000,000	hundred billions	700,000,000,000
7.	5,000,000	millions	5,000,000
8.	5,000	thousands	5,000

Write the digit that is in each place in the number 793,462,105,800.

9. tens __0__
10. thousands __5__
11. hundred thousands __1__
12. millions __2__
13. billions __3__
14. ten millions __6__
15. hundred billions __7__
16. hundred millions __4__
17. hundreds __8__
18. ten billions __9__

Objective: To order larger numbers.

Materials: Overhead spinner or 10-sided dice, base ten blocks, playing cards

Introductory Activities

Comparing Numbers

Ask the class what it means to put a set of objects in order. (to arrange them from left to right according to some rule)

Choose 3 to 5 students, and ask them to stand in front of the class. Ask volunteers to suggest ways to order the students. Some possibilities are by age, height, or alphabetically by name.

We can also order and compare numbers. Put base ten blocks (tens and ones) in a pile in the middle of each small group. Students take turns picking up a handful of blocks and making a number with the blocks. Students compare numbers and the person with the greatest number wins one point.

Write any two different 2-digit numbers on the board, e.g., 37 and 52. **Build each number with blocks. Which is more?** (52) **How do you know?** (It has more units blocks or because the digit in the tens place is greater.) **The greater than and less than signs must always open to the bigger, greater or larger number. How would we draw the symbol between these two numbers?** (37 < 52) **To read number sentences, we always begin at the left, just as when reading stories. How would we read this number sentence?** (37 is less than 52)

Repeat with other 2- and 3-digit numbers.

About This Page

Carefully review the pattern for comparing numbers in the illustration. Ask the students to identify the first place value where the digits are different and have them compare those digits.

Follow Up Activities

Comparison Game

Game for 2–4 players: Each group should have a deck of 36 cards (tens and face cards removed). Deal 3 cards to each player. Each player then arranges the cards in his hand to form a 3-digit number. Players state whether they think their number will be the highest, lowest, or middle

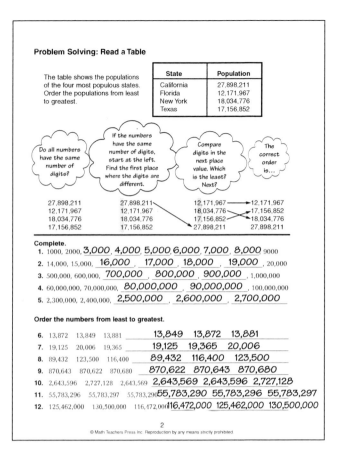

number when compared with the other players. All players show their hands and each player who guessed correctly wins a point.

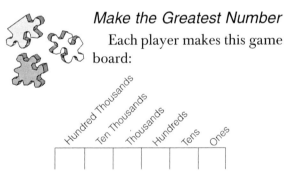

Make the Greatest Number

Each player makes this game board:

Use an overhead spinner divided into 10 parts labeled 0–9 (or a 10-sided die). **I will call out six different numbers. Each time, you must write the number in one of the 6 boxes. You cannot change the number once it is written down. The object of this game is to think about how to place the numbers, so you will write the greatest number.** Each player who has written the greatest number receives 1 point. Play several rounds. After each round, have students order the list of numbers for each round from least to greatest.

Skill Builders 2-1

Objective: To round numbers to the nearest 1000 using models and a halfway number.

Materials: Base ten blocks, playing cards, Place Value Mat (Master 10 and 11)

Vocabulary: rounding

Introductory Activities

Rounding to the Nearest 10

Give examples of times when students might need to approximate a number instead of an exact number.

If there are 28 students and you wanted to know *about how many* students there are, what would your answer be?

Have students use base ten blocks. **Build the number 28.** (2 tens and 8 ones blocks) **We use rounded numbers to find about how much a number is. About how many tens would you say 28 is? Build 2 tens and 3 tens. Is 28 closer to 20 or 30?** (30) **How do you know?** (It takes 2 ones blocks to get to 30, but 8 blocks to get to 20.) **Is 28 closer to 2 tens or 3 tens?** (3 tens)

Repeat with 21, 26, 25. Students may discover that numbers above the halfway number 25 are rounded up and numbers below 25 are rounded down. **The halfway number, 25, is rounded up by agreement. It is fair because 5 numbers that have 2 in the tens place (20–24) round down and 5 numbers (25–29) round up.**

Rounding to Nearest 100 and 1000

Practice these steps to round to the nearest 100.
1. Build the number with base ten blocks.
2. Skip count by 100s and place the number between two groups of 100s.
3. Think of the halfway number between the two groups of 100.
4. If the number is above or equal to the halfway number, round it up to the next hundred. If it is below, round it down to the lower hundred.

To round to the nearest 1000, count by 1000s to 10,000. Place the number between two groups of thousands (place 5716 between 5000 and 6000). Decide if the number is above or below the halfway number (5716 is above the halfway number of 5500).

About This Page

Build the number 1843 using base ten blocks. **Skip count by 1000s. Between which group of 1000 does 1843 come?** (1000 and 2000) **What is**

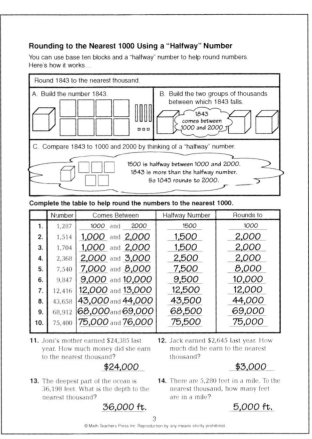

the halfway number between 1000 and 2000? (1500) **Is 1843 above or below this halfway number?** (above) **Is 1843 closer to 1000 or 2000?** (2000) Repeat with 1649 and 1532.

Follow Up Activities

Rounding Relays Game

Divide students into groups of equal size. Give each group a deck of 36 playing cards (tens and face cards removed).

The first player on each team turns over four cards and writes a 4-digit number on the chalkboard. The second player on each team rounds the number to the nearest 1000. If the player rounds the number correctly, then she removes the four cards from the deck and turns over the next four cards to form another 4-digit number. If the number is not rounded correctly, the cards are returned to the pile before the player selects four more cards. The first team to get rid of their 36-card deck is the winner.

Journal Prompt

Explain how to round 2368 to the nearest thousand. Use the words **between** and **halfway** number in your explanation.

Skill Builders 3-1, 3-2

Objective: To understand the order or commutative property.

Materials: Square tiles (or squares cut from Master 7), empty egg cartons, multiplication chart (made from Master 7)

Vocabulary: property, commutative (order) property, multiplication

Skill Builders Vocabulary: associative (grouping) property

Introductory Activities

Using Models

Pretend that we are lining up with a partner to go to lunch. Call one set of partners at a time to line up facing the door until there are six groups of two each. **How many groups of two are lined up?** (6 groups of 2) **How many total students?** (12) **What multiplication fact is shown?** ($6 \times 2 = 12$)

Now, let us have these students facing the door turn so they face the wall on their right. Now there are 2 rows or groups of 6. **How many total students?** (12) **What multiplication fact is shown?** ($2 \times 6 = 12$) **How does the total number of students compare? Why?** (The total is the same because the students have only changed their position.)

Write on the chalkboard or overhead:
$$6 \times 2 = 2 \times 6$$
$$12 = 12$$

We say that the order in which we multiply two numbers does not change the product. This is called the *order* or *commutative property*. A property is a rule that is always true.

You may use empty egg cartons cut into groups of two to demonstrate other examples of partners lining up, e.g., 5 groups of 2 egg cartons are turned to become 2 groups of 5.

Each time write the 2 facts on the board:
$$5 \times 2 = 2 \times 5$$
$$10 = 10$$

We know that you can multiply the numbers in any order and the product is the same. Is this true for the operations of addition, subtraction and division as well?

Is the addition of two numbers such as 6 and 2 commutative? (yes) Give an example to illustrate. ($6 + 2 = 2 + 6$) **Is the subtraction of two numbers commutative?** (no)

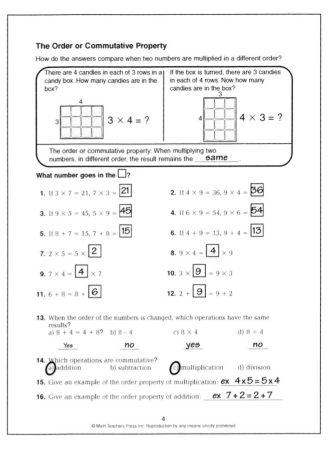

Give an example to illustrate. ($6 - 2 \neq 2 - 6$)
Is the division of two numbers commutative? (no) Give an example to illustrate. ($6 \div 2 \neq 2 \div 6$)

About This Page

Use square tiles (or squares cut from Master 7) to demonstrate several examples.

Follow Up Activities

Patterns with the Multiplication Facts

Give each student a multiplication table or have them complete one. Have students draw a box around all the square numbers (1, 4, 9, 16, 25, 36, 49, 64, 81). **What kind of line is drawn through these numbers?** (a diagonal line) **Fold your table on the diagonal and hold it up to the light. What do you see?** (like products match in position, e.g., 6×8 matches 8×6)

Journal Prompt

Use an example to explain the commutative or order property. Draw a picture, use words and symbols.

Skill Builders 5-1, 5-2

Objective: To add numbers of uneven length in horizontal format.

Materials: Place Value Mat (Masters 10 and 11), base ten blocks, Centimeter Graph Paper (Master 2), 100 Basic Addition Facts (Master 3)

Introductory Activities

Timed Test on Addition Facts

Use Master 3 to assess mastery of the 100 basic addition facts. If the tests are to be timed, use the following instructions. **Today we are going to take a timed test on the 100 addition facts, to see how many you can remember in 3 minutes. You may not be able to complete all the 100 facts in 3 minutes, but we will periodically give the test, and you will gradually see an increase in the number you are able to complete. When I say "begin," begin. When I say "stop," stop.**

Record the number correct. Continue administering the test on a regular basis. Have students record their results on a line graph, so they can see their gradual improvement. Analyze frequently missed facts and suggest strategies or practice for remembering the fact.

Write 45 + 312 + 6 horizontally on the chalkboard. Have students build each number and join blocks on a Place Value Mat. **How many ones?** (13) **Can you exchange?** (Yes, 10 ones for 1 ten.) **How many tens?** (6) **Can you exchange?** (No) **How many hundreds?** (3) **Can you exchange?** (No) **What number is shown?** (3 hundreds 6 tens 3 ones or 363)

Draw 3 columns labeled ones, tens and hundreds on the chalkboard. Copy the digits from each number under the correct column. **When we recopy these numbers from horizontal to vertical format, the digits must always be lined up starting at which side?** (At the right because every number will have a digit in the ones place.)

About This Page

Use a transparency of Centimeter Graph Paper (Master 2) to demonstrate how students should rewrite each problem correctly in vertical format. Use of the graph paper will reduce careless errors made by many students as they recopy the problem in vertical format.

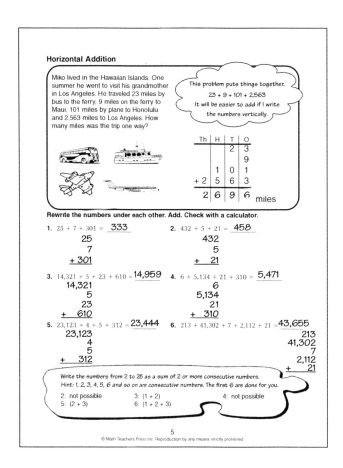

Follow Up Activities

Finding the Pattern

A young German by the name of Carl Gauss found a pattern to find the sum of the numbers from 1 to 100. Can you find the pattern? Hint: How many different pairs of whole numbers have 100 as their sum? Allow time for exploration.

Here is one solution to the problem:

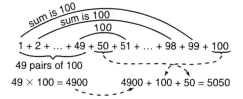

Skill Builders 6-1

Objective: To subtract numbers of uneven length in horizontal format.

Materials: Base ten blocks and Place Value Mats (Masters 10 and 11), calculators (optional), 100 Basic Subtraction Facts (Master 4)

Introductory Activities

Timed Tests on 100 Subtraction Facts

Use Basic subtraction Facts (Master 4) to assess mastery of the 100 subtraction facts. Three-minute, timed tests may be given periodically to increase the number correct within a given time period.

Uneven Subtraction

Make up a problem related to 2178 – 46.
Write on the board in horizontal format:
2178 – 46

Build 2178 on your mat. Remove 6 ones. Remove 4 tens. What number is left? (2132) Show how to rewrite the problem in vertical format on the chalkboard.

```
  2178
-   46
  2132
```

Discuss how the digits in the smaller number must be carefully aligned with digits of the same place value in the larger number.

Repeat the activity with 1246 – 85, where 1 thousand must be exchanged for 10 hundreds and 1 hundred exchanged for 10 tens before the subtraction can take place.

Checking Subtraction

Write on the board: There are 9 boys and 7 girls in the computer class. How many members are in the class?

Use base ten blocks on a whole-part mat to aid visualization of the whole-part relationship in addition and subtraction.

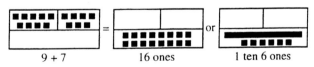

9 + 7   16 ones   1 ten 6 ones

In addition we put parts together to make the whole: part + part = whole.

Write on the board: The band has 16 members. Seven of them are girls. How many are boys?

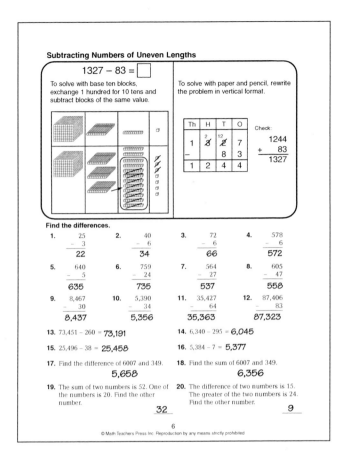

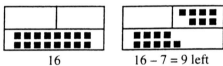

16   16 – 7 = 9 left

In subtraction, we take one part from the whole to see what is left: whole – part = part.

Addition and subtraction are opposites. Write 2 addition facts and 2 subtraction facts using these numbers. (7 + 9 = 16, 9 + 7 = 16, 16 – 9 = 7, 16 – 7 = 9) **Because addition and subtraction are opposites, we can check subtraction by addition.**

Write on the board: Bill has $84. He spent $29 to buy a video. How much does he have left?

Use base ten blocks on a Place Value Mat to solve the problem. Use pencil and paper to check.

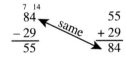

About This Page

Direct attention to the example at the top of the page. Use base ten blocks to demonstrate the regrouping of 1 hundred for 10 tens and how this trade is recorded.

Skill Builders 7-1

6   6E Teacher Manual

Objective: To estimate sums of larger numbers by rounding to the leading digit.

Materials: Base ten blocks, 10-sided dice, calculators (optional)

Vocabulary: leading digit, estimate

Introductory Activities

Estimating Addition

Approximate numbers are often used to find *about* how much the answer is. We round each number to the digit in the greatest place value and then use these leading digits to estimate an answer.

Make up a problem related to the estimation of the sum of 57 + 32.

Example: You have 57¢ and 32¢. About how much money do you have?

Ask students to round each number to the leading digit and then find the sum of the rounded numbers (60 + 30 = 90).

Repeat for estimating the sum of two 3-digit numbers, e.g., 837 + 458 (800 + 500 = 1300).

About This Page

Read the example at the top of the page. Emphasize that students will not have time to solve these problems with paper and pencil.

Allow students 3 or 4 minutes to estimate the sums for problems 1–12. Have students explain their estimates.

Follow Up Activities

For problems 13–16, students may find the actual answer using paper and pencil or calculator and then compare the actual answers to the estimates.

Estimating Sums

Small group activity. Each player makes a score sheet to record the number thrown on the die.

	hundreds	tens	ones
Throw 1			
2			
3			
4			
5			
Total			

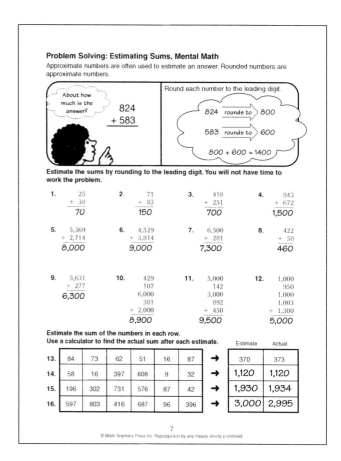

Players take turns rolling a 10-sided die three times. All players record the number thrown on their score sheet in the hundreds, tens or ones place. After five turns (of rolling the die three times each turn), the players find the sum of the number on their cards. The player closest to 1000 without going over wins.

Journal Prompt

Complete these sentences:
Rounding numbers to the leading digit when adding is helpful because … .
In real life I would use this method when … .

Skill Builders 49-1

6E Teacher Manual 7

Objective: To estimate differences by rounding to the leading digit.

Materials: playing cards, calculator (optional)

Introductory Activities

Estimating Subtraction

Write a word problem related to the estimation of 725 − 481. Ask students to identify the place value of the leading digit (hundreds), round each number to the nearest hundred and subtract the rounded numbers (700 − 500 = 200).

What is your estimate? (200)

Repeat with two 4-digit numbers, e.g., 6384 − 1542 (6000 − 2000 = 4000).

About This Page

Emphasize that students will not have time to solve their problems with paper and pencil.

Allow students 3 or 4 minutes to estimate the difference for problems 1–14. Have students explain their estimates. For problems 15–18, estimates may vary.

Follow Up Activities

Have students find the actual answer using paper and pencil or calculator. Compare the actual answer to the estimated answer.

Estimating Sums and Differences Relay

Divide the class into teams of equal size, each with its own 36-card team deck (tens and face cards removed). The first player on each relay team turns over 6 cards and writes a 3-digit addition problem on the board using all 6 numbers, e.g.,

942
+ 316

The second player on each team rounds each addend to the leading digit and finds the estimated sum:

900
+ 300
1200

If the second player rounds and estimates the sum correctly, the 6 cards are removed from the team deck, and the second player turns over

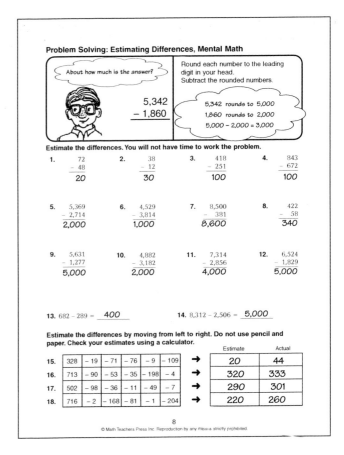

6 new cards to make another problem for the third player to round and add.

If the second player does not round and estimate correctly, the cards are returned to the team deck, and the player selects six new cards to make an addition problem. The first team to remove all cards from the deck is the winner.

The game may be varied to make addition problems involving 4 digits plus 3 digits, 5 digits plus 4 digits, and so on.

Vary the game to estimate with subtraction.

Mental Math

Explain that you will present problems aloud, asking the students to determine in their heads whether the answer is more or less than a given number. If the answer is more, they are to put their thumbs up; if the answer is less, they are to put their thumbs down; if the answer equals the given number, they are to put their thumbs sideways.

Example: Is the sum more or less than 100?
18 + 71   (thumbs down)
18 + 92   (thumbs up)
18 + 82   (thumbs sideways)

Skill Builders 49-2

Objective: To use base ten blocks to multiply by multiples of ten. **DVD**

Materials: Base ten blocks, Place Value Mats (Masters 10 and 11), index cards, 100 Basic Multiplication Facts (Master 5)

Introductory Activities

Timed Tests on 100 Multiplication Facts

Begin weekly tests of the multiplication facts using Master 5. Allow 3 minutes and record the number correct on a bar or line graph so students see their improvement.

Real-World Math

Use real-world problems to begin your math lessons. Make up a problem related to 3 × 14.

Example: There are 14 colored pencils in each package. How many pencils are in 3 packages?

Read the problem together and decide that this is a multiplication problem. **How do you know this is a multiplication problem?** (We are putting together groups of equal size.) Ask a student to write a number sentence to solve the problem 3 × 14 =.

We can use base ten blocks to solve this problem. Build 3 × 14 (3 groups of 1 ten and 4 ones). **Put the ones together and the tens together. How many units or ones do you have?** (12) **How many tens blocks?** (3) **Can we make an exchange?** (Yes, we can exchange 10 ones for 1 ten.) **Now how many unit cubes do you have?** (2) **How many ten blocks?** (4) **How much is 3 groups of 14?** (4 tens 2 ones or 42)

Record the answer on the chalkboard as you discuss the exchange that results in regrouping or *carrying* the 1.

Multiplying Multiples of Ten

Demonstrate each of the following problems with base ten blocks:
1. 10 × 10 (10 tens blocks) = 1 hundred or 100
2. 20 × 10 (20 tens blocks) = 2 hundreds or 200
3. 30 × 10 (30 tens blocks) = 3 hundreds or 300
4. 20 × 20 (20 groups of 2 tens blocks) = 4 hundreds or 400

Describe the pattern for the product when you multiply multiples of ten. (Multiply the leading digits and write two zeros to the right.)

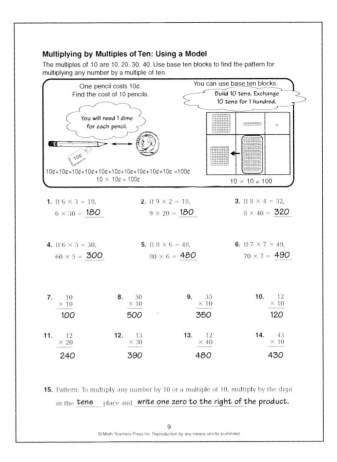

About This Page

Demonstrate the problem at the top of the page with base ten blocks. Have student volunteers demonstrate the solution of problems 1, 2 and 3 with base ten blocks.

For problem 1, 6 groups of 3 tens would be built and combined to become 18 tens; 18 tens would be changed to 1 hundred 8 tens or 180. It is important that students understand the pattern for multiplying by multiples of ten, as it is part of the pattern for multiplying by any 2-digit number.

Follow Up Activities

Cooperative Group Practice

Divide the class into groups of 2. Each group will need 9 index cards with one of the multiples of 10 written on each card: 10, 20, … 90. Turn the cards face down and select any 2 cards. Student 1 finds the product with base ten blocks; Student 2 finds the product with pencil and paper.

After 4 problems have been worked correctly, shuffle the cards and have the players reverse roles to solve 4 more problems.

Skill Builder 50-2

Objective: To multiply a 3-digit number by a 2-digit number.

Materials: Base ten blocks, Place Value Mats (Masters 10 and 11), calculator (optional)

Introductory Activities

Multiplying by 2-Digit Numbers

Write on the board: A box of footlong hot dogs contains 100 hot dogs on a string. Each hot dog is 12 inches long. How long is the string of hot dogs?

Ask a student volunteer to read and draw a picture of the problem. **This problem suggests 100 strips of 12 along a number line.**

What is the action in the problem? (putting together groups of equal size) **Which operation puts groups of the same size together?** (multiplication)

Write on the board:
```
   100
 ×  12
```

Ask a student volunteer to use base ten blocks to solve the problem. (12 one hundred blocks are put together; 10 one-hundred blocks are exchanged for 1 thousand, leaving 1 thousand block and 2 hundreds blocks, or 1200.) Emphasize that a digit in the tens place times a digit in the hundreds place results in a digit in the thousands place.

Write on the board:
Since 12 = 10 + 2,
```
  100
× 12        is the same as

  100              100
×   2     +      × 10
```

About This Page

Read the example at the top of the page. Ask a volunteer to describe the action and write a number sentence to solve the problem.

When you multiply a number by 12 you are really multiplying it by which two smaller numbers? (10 and 2) **How much is 2 times 140?**

Write on the board:
```
   140
 ×  12
   280    (2 × 140)
```

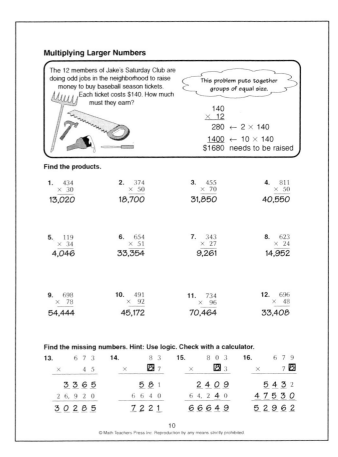

How much is 10 times 140? Show the answer with the fewest number of blocks. (1400)

Write on the board:
```
    140
 ×   12
    280    (2 × 140)
 + 1400    (10 × 140)
   1680    needed in all
```

Problems 13–16 may be solved using a combination of guess and check with logical thinking. Have students check their problems with a calculator.

Follow Up Activities

Journal Prompt

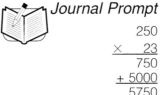

```
    250
 ×   23
    750
 + 5000
   5750
```

Prove the answer to be correct or incorrect. If it is incorrect, show how the answer should be corrected. Use words, diagrams and/or symbols to support your claim.

Skill Builders 8-1, 8-2

Objective: To divide up to a 6-digit number by a 1-digit number, zeros in the quotient. **DVD**

Materials: 90 Division Facts (Master 6), tennis balls, 10-sided dice, base ten blocks, paper plates

Vocabulary: dividend, division, divisor, quotient

Introductory Activities

Timed Test on Division Facts

Use Master 6 to administer tests on division facts. Allow 3 minutes if tests are to be timed.

Dividing with Base Ten Blocks

Make up real-world problems related to $2\overline{)2684}$ $2\overline{)3456}$ and $3\overline{)2214}$.

Use base ten blocks to build each dividend and share the blocks on a number of plates equal to the divisor.
Example: $2\overline{)2684}$

Build 2 thousands 6 hundreds 8 tens 4 ones and share the blocks, starting with the thousands blocks, on 2 paper plates.

About This Page

Build 3456 with blocks and share on 2 plates. Start sharing the thousands block first. How many thousands blocks on each plate? (1) **What should you do with the 1 thousands block left over?** (Trade 1 thousand block for 10 hundreds blocks.) **How many hundreds blocks go on each plate?** (7) **When you share 5 tens blocks on 2 plates, how many tens blocks go on each plate?** (2) **What shall we do with the 1 ten left over?** (Exchange 1 ten for 10 ones.) **How many ones go on each plate?** (8) **What number is on each plate?** (1728) **The steps in long division (Divide, Multiply, Subtract, Bring Down) are used each time a new block is shared.**

Have students use blocks and plates to divide the first 3 problems. Then have them use the 4 steps in the division pattern to complete the page.

Follow Up Activities

Opposite Ball Game

Divide students into small groups of 4 to 6 students. Each group forms a circle with one player in the middle. The middle player throws a tennis ball to anyone in the outer circle, saying a multiplication fact at the same time, e.g., "8 × 6

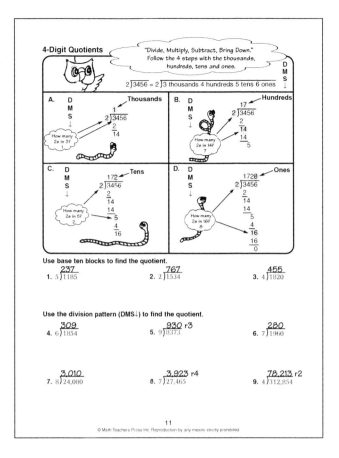

equals ... ." The player is to catch the ball, say "48" aloud, and throw the ball back to the center, saying a related division fact at the same time. "48 divided by 6 equals ... ." If a person misses the answer, or is unable to restate the opposite operation, she sits down. Play continues until only one player is standing in each group.

Smallest Quotient Game

Divide the class into groups of 2 to 4 students. Each group will need three 10-sided dice. Players take turns rolling all 3 dice and making a division problem using two of the numbers rolled as a 2-digit dividend and the other number as the divisor, and then finding the quotient. Zero may not be used as the first number of the dividend nor as the divisor.
Example: $4\overline{)76}$

After each round, the player with the smallest quotient is the winner. Continue for several rounds.

Many variations of the game may be played: the largest quotient wins, the smallest remainder wins, etc.

Skill Builders 9-1

Objective: To divide by multiples of ten.

Materials: Base ten blocks, paper plates, multiplication tables (made from Master 7)

Introductory Activities

Discover the Pattern

There are two models for division, the sharing model and the subtraction model. Write the following problems in a table.

Division	Quotient
50 ÷ 10	
100 ÷ 10	
200 ÷ 10	
80 ÷ 20	
100 ÷ 20	

Demonstrate each method and ask which method is easier. (In these examples, subtraction is easier because you must use 10 or 20 paper plates to use the sharing model.)

For the first example, they could share 50 ones on 10 paper plates (sharing model) or remove groups of 10 from 50 (subtraction model). Students will see that the subtraction model is more efficient for these problems. After finding the quotients, have students study the results to find the pattern for dividing by multiples of 10. (First, look at the number of zeros in the divisor; cross out all the zeros in the divisor and the same number of zeros in the dividend. Then divide the remaining leading digits.)

Example: 200 ÷ 10 = 20

Write on the board: A large bag contains 160 balloons. They are to be divided into 20 small bags. How many balloons in each bag?

Have a volunteer read the problem and decide on an operation (division).

Have another student build 160 with base ten blocks (1 hundred, 6 tens). Ask which model will be easier to solve the problem and explain why. (It will be easier to remove groups of 20 than to share 160 units on 20 paper plates.)

We can also solve this problem by using the multiplication chart. When dividing 160 by 20, refer to the multiples in the 2 row. Find the greatest number of multiples that could be subtracted from 16. (8)

About This Page

Review the 3 different ways to solve the problem at the top of the page.

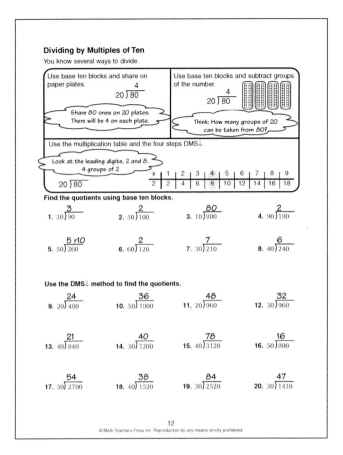

Can you share 1 hundred block on 60 plates? (no) Write 0 above the 1. Can we share 12 tens blocks with 60 plates? (no) Write 0 above the 2. Can we share 120 one blocks on 60 plates? (Yes, there will be 2 on each plate.)

At first, you may wish to have students write zeros in the front of the quotient each time the divisor does not go into the dividend to aid this process. Working the problem on graph paper or drawing lines between places will help them align digits.

Example:

$$60 \overline{)\, 1\,2\,0 }^{\,0\,0\,2}$$
$$1\,2\,0$$

It may also be helpful to have students underline the digits one at a time that will be needed to obtain the first non-zero digit in the quotient.

Example:

$$4 \overline{)1600}^{\,400}$$

Follow Up Activities

Skill Builders 10-1

Objective: To use base ten blocks to divide.

Materials: Base ten blocks, Place Value Mats (Masters 10 and 11), paper plates

Introductory Activities

Different Ways to Divide

Write on the board: You are packing eggs in cartons that hold 1 dozen each. How many cartons are needed to pack 96 eggs?

Ask a volunteer to read the problem, identify the question, the needed facts, decide on the process and write the problem on the board:

12)96

How many eggs are there in all? (96)
How many are to be put in each group? (12)
Build the big number, 96, on your Place Value Mat. Remove groups of 12. How many groups of 12 can be removed altogether? (8)

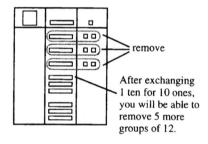

Record the solution: 8
                     12)96
                     96 total number removed

Write on the board:
   Sixteen students are sharing $192.
   How much will each person receive?

Ask a volunteer to read the problem aloud, find the question, needed facts, decide on a process and write the problem on the board:

16)192

How many dollars are there in all? ($192)
How many students are to share the money? (16) **Build the big number.** (192)
Count out 16 plates. Share the big number on each plate. Can you share 1 hundred block on 16 plates? (No, we must first trade 1 hundred for 10 tens.) **How many tens go on each plate?** (1 ten) **How many tens did we share?** (16) **How many are left to share?** (3 tens) **Can you share 3 tens on 16 plates?** (No, we must trade 3 tens for 30 ones.) **When you share 32 ones on 16 plates, how many ones

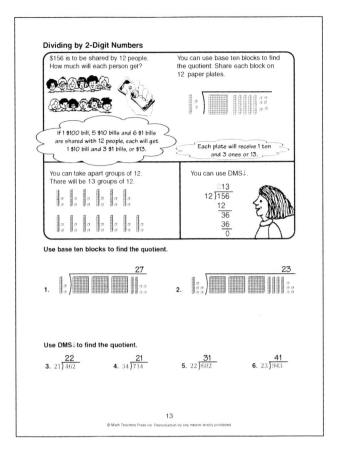

are on each plate?** (2) **How many blocks are on each plate?** (1 ten 2 ones or 12)

Have another student describe how the blocks were shared on the plate and how the four steps in division, DMS↓, are used to record the problem with paper and pencil.

```
        012
   16 )192
        16
        32
        32
```

About This Page

Read the illustration together. Use base ten blocks to model the different ways to solve the problem.

Follow Up Activities

Skill Builders 10-2, 10-3

Objective: To read a table. To estimate products by rounding.

Materials: Base ten blocks, calculators, playing cards

Introductory Activities

Rounding Patterns

1. Build the number with base ten blocks.
2. Skip count by 100s and place the number between two groups of 100s.
3. Think of the halfway number between the two groups of 100.
4. If the number to be rounded is above or equal to the halfway number, round it up to the next hundred. If it is below the halfway number, round it down to the lower hundred.

Estimating Products

After students are able to round to the nearest ten, hundred or thousand, have them use the pattern of the zeros to estimate the products. Write on the board:

	Rounds to	Estimate
81 × 8		
48 × 4		
641 × 4		
809 × 7		
23 × 81		
97 × 43		
333 × 68		

Have students copy the table and circle the leading digits in the factors.

About This Page

Read the problem at the top of the page. Ask questions about the table. **How far is it from Northville to Elmira? How far is the round trip?** Ask similar questions about problems 1–4. If a calculator is not available, students will use paper and pencil to find the product of the two numbers selected.

Follow Up Activities

Estimating Products Relay

Divide the class into teams of equal size. Give each team a deck of 36 cards (tens and face cards removed). At a signal, the first person on each team turns over the top four cards and writes any 2-digit times a 2-digit problem on the board which uses the numbers on the cards, e.g., 4, 1, 2, 7 could be 12 × 47 or 41 × 27, etc.

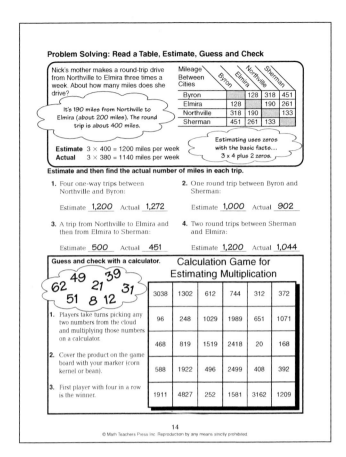

The second player then rounds the numbers and writes an estimate for the product on the chalkboard. If the rounding and estimation were done correctly, the 4 cards are removed from the team pile and the second player selects four more cards and writes a 2-digit times 2-digit problem for the third player to round and estimate. If the problem is not estimated correctly, the cards go back in the team pile. The first team to get rid of all their team cards is the winner.

Greatest Product Variation

Ask students to arrange digits so they have the greatest product. After each team has completed nine problems, find the sum of the nine products to determine the winner.

Journal Prompt

Explain how to estimate an answer to multiply 641 × 9. Why is estimating products an important skill to learn?

Skill Builders 50-1

Objective: To estimate quotients by rounding.

Materials: Base ten blocks, index cards, paper plates

Introductory Activities

Discovering the Pattern

Write on the board: 60 ÷ 30 = ?
Demonstrate the two methods of division:
1. Sharing 60 on 30 paper plates (each plate will have 2 ones) and
2. Removing the divisor (30) from the dividend (60) until there are none left. (2 groups of 30 may be removed)

Which method seems easier for this problem? Why? (The second method is quicker.)
Write the following table on the board.
Use base ten blocks to complete the table. Ask students to look for a pattern for dividing by multiples of ten and write the pattern on their paper.

Divide	Quotient
600 ÷ 30	
280 ÷ 70	
320 ÷ 80	
800 ÷ 20	

Pattern: Cross out equal numbers of zeros in the dividend and divisor. Then divide the leading digits. Example: 600 ÷ 30 becomes 60 ÷ 3 or 20.

About This Page

Complete the table together at the top of the page. Ask several students to describe the pattern they found. (First, I crossed out the same number of zeros in the dividend and divisor. Then I divided the numbers left. If there were any remaining zeros, they were written in the quotient.)

Look at several problems together, asking students to explain how they would round the dividend and divisor in the problem. Then tell the students you will allow them 3 or 4 minutes to estimate the quotients, emphasizing that they will not have time to find the actual answer.

After the specified time, have students explain their estimates. There may be more than one good estimate for each problem.

Have students find the actual answers and then compare their actual answers to the estimates.

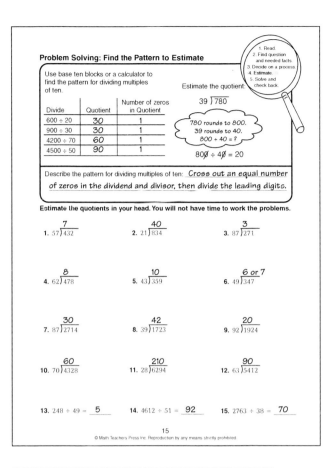

Follow Up Activities

Mental Math

Explain that you will present problems aloud, asking the students to determine in their heads whether the answer is more than or less than a given number. Have students use their thumbs to indicate if the answer is more (thumbs up), less (thumbs down) or equal to (thumbs sideways) the given number.

Example: Is the quotient more than, less than or equal to 50?

50 ÷ 10   (thumbs down)
420 ÷ 20 (thumbs up)
500 ÷ 10 (thumbs sideways)

Objective: To introduce a five-step problem solving model.

Materials: Posterboard for making the five-step problem-solving model classroom chart, index cards

Introductory Activities

Steps in Problem Solving

Write on the board: 256 children from Kennedy School did their homework on Saturday. 472 children did their homework on Sunday. How many more children did their homework on Sunday?

Today we are going to learn some steps to help solve word problems. What is the first thing we must do to solve the problem on the board? (Read it.) Ask a student to read the problem aloud and tell the story in her own words.

Write on the board or on a large poster:
Steps in Problem Solving
Step 1. Read and Understand.
What is the next step in solving this problem? (Find the question and the needed facts) **What is the question?** Underline the question, How many more children did their homework on Sunday?
What are the facts? (256 children did homework Saturday; 472 children did homework Sunday.) Circle these facts on the board.

Write on the board:
Step 2. Find the question and the needed facts.
What is the next thing you must do? (You must decide how to solve the problem.)

Write on the board:
Step 3. Decide on a process.
Ask students what strategy to use. Elicit strategies such as act out the problem, draw a picture and use a model or make the numbers in the problem smaller. **What process should we use to solve the problem?** (subtraction)

Before we actually solve the problem, it is a good idea to guess or estimate the answer. Then you will know if your actual answer is reasonable.

Write on the board:
Step 4. Guess or Estimate.
Ask a student volunteer to estimate for Step 4 (500 − 250 = 250). **After estimating the problem what should we do?** (Solve the problem.) Have a student volunteer solve the

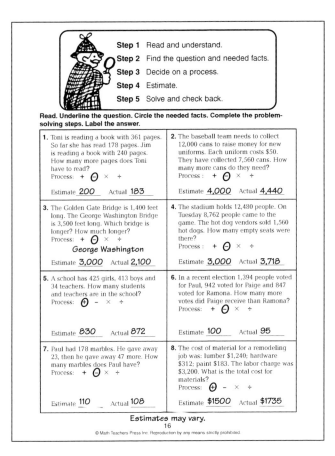

problem. **How can we decide if the answer is correct?** (Check the problem by putting the answer back in the problem)

Write on the board:
Step 5. Solve and check back.

About This Page

Read the first problem together. Ask a student to retell the story and restate the question. Have students underline the question. Ask a student to list the facts while you write each fact on the board. **Which facts are needed to answer the question? Are there any unnecessary facts we can cross out?** (Yes, Jim's book has 240 pages.)

Students may complete the page on their own or with a partner.

Review the structure that addition and subtraction are reversible operations.

What is the action in this problem? (part of the book is read) **Which operation should we use?** (subtraction)

Write on the board: 361 − 178 = N
Ask students to estimate, then solve the problem.

Follow Up Activities

Skill Builders 45-1

Objective: To use a variety of problem-solving strategies to solve a word problem.

Materials: Posterboard for making the Strategies to Solve Word Problems classroom chart

Introductory Activities

Strategies for Solving Problems

Write on the board: There are 10 chapters in a book and 25 pages in each chapter. Joyce has read more than 100 pages. How many pages are in the book?

Refer to the five steps in problem solving chart:
1. Have a student volunteer read the problem aloud and then retell the story.
2. Have a student volunteer underline the question and circle the needed facts.
3. Ask whether each of the following five strategies might be used to solve the problem.
   a) Could we act out the problem? (Yes, students could look at an actual book with a given number of chapters and a given number of pages to act out the problem.)
   b) Use a model? (Yes, students could build 10 groups of 25 with base ten blocks, put like blocks together and record the answer.)
   c) Draw a picture? (Yes, students could draw a sketch of 10 chapters with 25 pages in each to visualize the problem as putting together groups of equal size.)
   d) Simplify? (Yes, the numbers could be changed to 10 and 20.)
   e) Make a table? (Yes, the table would be as follows):

Chapters	Pages
1 chapter	25 pages
2 chapters	50 pages
10 chapters	250 pages

4. Estimate.
5. Solve and check back.

Complete the last two steps by having a student estimate an answer, solve the problem and compare the actual answer to the estimate.

Write the five commonly used strategies on a chart titled "Problem-Solving Strategies."

About This Page

Read the list of strategies and the first problem together. By asking students which of the strategies they might use, they become aware that a variety of strategies may be used to solve any problem.

Can we "Act It Out" to solve this problem? (Yes, we could walk a given number of miles for a given number of hours.)

Can we use a model? (Yes, we could build 16 groups of 58.)

Can we draw a picture? (Yes, we could sketch the problem as 16 groups of 58 miles each on a long number line.)

Can we simplify? (Yes, we could use simpler numbers such as 6 miles and 2 hours.)

Can we make a table? (Yes. 1 hour = 58 miles; 2 hours = 116 miles, and so on.)

Although we can use all five strategies, which would you probably prefer to use for this problem? Why? (Allow students the chance to express their reasons for preferring one strategy over the other.)

Follow Up Activities

Skill Builders 45-2, 45-5

Objective: To find the average of a set of numbers.

Materials: Counters, box of animal crackers, paper plates, 6-sided dice

Vocabulary: average, mean

Introductory Activities

Sharing Animal Crackers

Purchase a box of animal crackers and count the number of crackers in the box. If necessary, remove crackers so that the number of crackers is divisible by 6.
What if six friends were going to share the crackers inside the box? Would this be a fair way to share them? (Open the box and divide the cookies into six groups of unequal size.)
Why or why not? Can anyone think of a fairer way to share the cookies? Discuss with your partner how you think they might be shared more fairly and how many each person should get.
After a few minutes, ask one or two volunteers to describe a fairer method and how they found the number each person should get. (I added up all the cookies and then divided by 6.)
This is what you do when you want to find an average or number representative of the group. Another name for average is mean.

Making Up Problems

Distribute 25 to 30 counters to each small group. **Pretend that these counters are cookies or other objects you wish to share fairly. Make up a problem involving averages. You do not have to use all the cubes.**
Collect the problems and exchange them among the students to solve.

About This Page

Read the example together at the top of the page. Have students work problems 1–4 with a partner with each person working one of the two steps (add the numbers and divide by the number in the set).
Have students complete the page on their own.

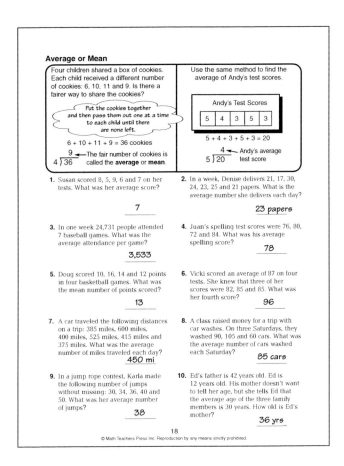

Follow Up Activities

Average Dice

Each small group of 2–4 students will need a six-sided die. Each player rolls the die five times, writing down each number as it is rolled.
Players calculate the average of their five numbers. The player with the higher average is the winner.
In a variation of the game, two players decide how many rolls the other player will have. The first player rolls the die no more than 5 times; the second player tells the first player to stop after 3 or 4 rolls or may allow the player to roll 5 times. The first player calculates the average. The players reverse rolls. The player with the higher average is the winner. Play several rounds.

Journal Prompt

Clark's test scores were 59, 78 and 49. What was his average score? Solve the problem. Then explain the steps you used to find the average.

Skill Builders 46-1, 46-2

Objective: To collect data and organize in a table. To find the mean, range, mode.

Materials: Index cards, calculators

Vocabulary: data, mean, range

Introductory Activities

Brainstorming Topics

Explain to the class that they are going to be collecting data about themselves and then organizing the information in a meaningful way. Begin suggesting topics that might be considered and record each on the board. Choose topics whose data is numeric.

Possible data to be collected:
- Height
- Number of letters in first name
- Number of pets in the family
- Number of books read each month
- Number of hours watching TV per week
- Number of hours exercising per week
- Number of hours sleeping each night

Ask the class to suggest topics they think are interesting. Then have each class member write down the five most interesting topics, tally the selections and collect data on the five most interesting topics.

Collecting Data

Divide the class into four groups. Give each group a pile of index cards. Announce the topic on which data is to be collected, e.g., number of hours watching TV per week. Have each student write a sentence on their index card, e.g., "I watch 8 hours of TV each week." Then have each group find the average number of hours for the whole group.

Discuss the small group average for each measurement, paying special attention to the meaning of the remainders. Could there be an average of $5\frac{2}{5}$ hours spent each week? Decide what should be done with the remainder.

After checking the average for each small group, have each student predict the average number of hours for the class. Collect the cards and ask what is the best way to find the average of a large group of numbers (use a calculator).

Write the number from each card on the board or overhead. **Which is the least number of**

hours? the most? We say that the range of the data is from the lowest number to the highest number.

Using a calculator or overhead calculator, add each number and divide by the number of students. Ask students if they had predicted close to the average or not.

About This Page

Divide the class into four groups. Each student will need four separate index cards. Announce the 4 topics and have students write each topic at the top of the 4 rectangular spaces. Then have each student complete 1 index card for each topic.

Have students find the range and average of each of the four topics in their groups. Collect all the cards and assign each group the task of finding the range and class average for one complete set of index cards. Each group will need a calculator.

After the class averages have been found, discuss the meaning of the remainders, e.g., could there be an average of $1\frac{1}{4}$ books read per student? $2\frac{3}{4}$ pets per student? $57\frac{3}{4}$ inches per student?

6E Teacher Manual

Objective: To name fractions from fraction bars. To identify similarities and differences among fraction bars. **DVD**

Materials: Fraction Bars®, overhead Fraction Bars® (optional)

Vocabulary: similarities, differences

Introductory Activities

Fraction Similarities and Differences

Distribute a set of fraction bars to each group of 2–5 students.

Each fraction bar in this set represents one whole or one unit such as one whole cracker or one whole brownie. Look through your set of fraction bars with your group. Discuss in what ways your bars are all alike and in what ways they are different. Record your findings in a table with two columns headed Similarities and Differences.

After 5 minutes, ask volunteers from each group to suggest the similarities and differences they have found.

Similarities	Differences
same size	colors
same shape	divided into different parts
congruent	number of shaded parts differs
same width and height	number of bars of any one color differs
same area and perimeter	
same thickness	
same weight	
same material	
all divided into parts of equal size*	

*It is very important that the last similarity–that each whole bar is divided into parts of equal size–be verbalized because this is the essential concept of a fraction.

Naming and Identifying Fraction Bars

The following three activities will assess students' abilities to translate from a concrete model to a written or spoken name.

1. Display overhead fraction bars. Ask a volunteer to name the shaded part of each bar. To encourage students to visualize the number of parts the whole is divided into, ask three questions: **How many parts has the whole been divided into? How many parts are shaded? What fractional part is shaded?**
2. Write a fraction on the chalkboard or overhead, e.g., ³/₄. Ask students to find a bar to match the fraction and draw a picture of the fraction. Repeat with other bars.
3. Say a fraction name aloud, e.g., ⁷/₁₂. Ask students to find a bar to match the fraction or draw a picture of the fraction. Repeat with other fraction names.

Problem Solving: Find the Pattern

These fraction bars have been sorted into groups by some way they are alike, or similar. Find the similarity.

1. Similarity: Divided into 2 parts
2. Similarity: 1 part shaded or fractions close to zero
3. Similarity: All parts shaded or equal to one whole
4. Similarity: One-half shaded
5. Similarity: One-third shaded
6. Similarity: Three-fourths shaded
7. Similarity: Less than one-half shaded
8. Similarity: All but 1 part shaded or close to one

What's My Secret?

With a partner or small group, students take turns selecting a subset of fraction bars that are alike in one way. Others in the group try to guess the secret. Demonstrate an example by showing all the bars of one color and have students guess the secret of the sorting. Other ways the students will sort may include everything shaded, nothing shaded, one part shaded, equivalent parts shaded, and so on.

About This Page

This page provides opportunities for students to generalize how three fractions are alike according to some attribute. Illustrate the first problem with overhead fraction bars.

Follow Up Activities

Skill Builders 11-1, 11-2

Objective: To identify fraction models as being in proper, improper or mixed-number form.

Materials: Two pounds of butter in ¼ pound sticks, apples or cardboard circles

Vocabulary: proper fraction, improper fraction, mixed number, numerator, denominator

Introductory Activities

Problem Solving: Using Models

Display a 1-pound carton of butter, showing the label "1 pound" on the outside of the carton. Open the carton and show the four ¼-pound sticks of butter inside. Ask students to say and write the name for 1 stick of butter (¼ lb.), 2 sticks of butter (²⁄₄ or ½ lb.), 3 sticks of butter (¾ lb.) and 4 sticks of butter (⁴⁄₄ or 1 lb.) Put the four ¼-pound sticks back into the carton to emphasize that ⁴⁄₄ is the same as 1 whole.

Display a second pound of butter and the ¼ pound sticks inside. Ask students to name and write the value of 1 full carton and 1 stick more. (⁵⁄₄ or 1¼), 1 carton and 2 sticks (⁶⁄₄ or 1²⁄₄ or 1½), 1 carton and 3 sticks (⁷⁄₄ or 1¾), 1 carton and 4 sticks (⁸⁄₄ or 1⁴⁄₄ or 2).

How does the amount of butter in ⁵⁄₄ compare to 1¼? (same amount) **The same amount of butter can be written in more than one form. Five-fourths (⁵⁄₄) is called an improper fraction. Why do you think it is called improper?** (It is more than 1 whole.)

1¼ is called a mixed number. Why do you think it is called a mixed number? (It involves a whole number and a fraction.)

Drawing Improper Fractions

Have students identify mixed numbers and improper fractions from fraction bars.
1. Display two blue fraction bars, ⁴⁄₄ and ¼. Ask students to write the value in two forms, as an improper fraction and as a mixed number (⁵⁄₄ and 1¼). Repeat with other examples.
2. Write improper fractions and mixed numbers on the board. Have students draw pictures or show fraction bars to match the numbers.
3. Say aloud, **one and one-fourth**. Have students draw pictures or show fraction bars to match. Repeat with ¹¹⁄₁₀, 1²⁄₃, and so on.

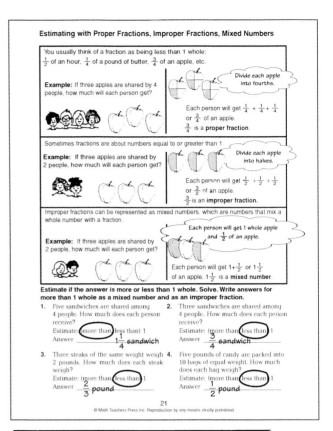

About This Page

Read and discuss the three examples. Use apples or cardboard circles on a drawing to demonstrate each example. Students will be able to find the patterns for interchanging mixed numbers and improper fractions, by studying the activities developed on these pages.

Follow Up Activities

Writing Word Problems

Have students write a word problem that will have a mixed number or an improper fraction as the answer. Collect and distribute to others to solve. Save for the class file.

Example: You and your friend each order a pizza. Each pizza is cut into 6 parts. After you eat a few pieces, there are 9 pieces left. what fraction is left? Write your answer in two different forms. (⁹⁄₆ or 1³⁄₆ or 1½)

Journal Prompt

What is the value of the fraction if the numerator is greater than the denominator? Give an example and explain, using words, symbols and diagrams.

Skill Builders 11-2

Objective: To interchange mixed numbers and improper fractions.

Materials: Real or play $1 bills (Master 8), quarters, dimes, overhead bills and coins

Introductory Activities

Problem Solving: Using Models

Use real, play or overhead bills and coins.

Display a $1 bill, asking how much it is worth. Point out the words ONE DOLLAR on the bill. Write on the board: $1 bill = 1 whole or 1 unit.

Display a quarter. Discuss the pictures on both sides and ask how much it is worth. Write on the board: 1 quarter = ¼ of 1 whole dollar.

Continue with a dime. (¹⁄₁₀ of 1 whole dollar)

Display 3 quarters. **What fractional part of a dollar is 3 quarters?** (³⁄₄) **Is 3 quarters an example of a proper, improper, whole number or mixed number?** (proper fraction)

Repeat with 4 quarters (an improper fraction or 1 whole); 5 quarters (⁵⁄₄ improper or 1¼ mixed); $1 bill and 3 quarters (1¾ mixed or ⁷⁄₄ improper).

Write on the board:
$$\tfrac{5}{4} = 1\tfrac{1}{4}$$
$$1\tfrac{3}{4} = \tfrac{7}{4}$$

Can you find a pattern to change the improper fraction ⁵⁄₄ to an equivalent mixed number? (Divide the numerator by the denominator)

Can you find a pattern to change the mixed number 1¾ to an equivalent improper fraction? (Multiply the denominator times the whole number, then add the numerator)

Have students write their patterns on a sheet of paper and draw a picture to explain why the patterns work.

About This Page

Read the examples at the top of the page. Be sure that each student writes out the patterns to complete the last two sentences. Ask a volunteer to explain out loud (or justify) why the patterns work. (When you divide 4 by 4, you are really finding the number of wholes, because there are ⁴⁄₄ in 1 whole. When you multiply the denominator by the whole number, you are really changing the whole number part into fourths because each whole has 4 fourths.)

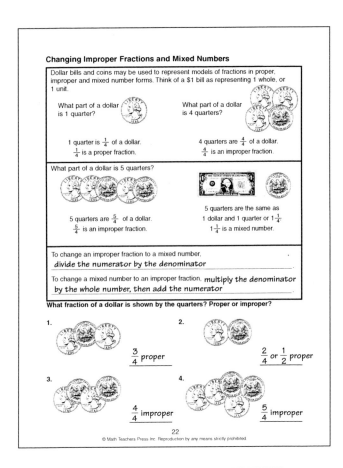

Follow Up Activities

Skill Builders 14-1

Objective: To use models to find the pattern for equivalent fractions. **DVD**

Materials: Fraction Bars®, chocolate bars scored into 12 pieces, interlocking cubes

Vocabulary: equivalence, equivalent fractions

Introductory Activities

Background

Many methods of renaming a fraction as an equivalent fraction need to be explored to fully develop the concept of proportionality and equivalent fractions. On these pages, students will use chocolate bars and interlocking cubes to write pairs of equivalent fractions as they are guided to discover this fundamental principle.

Equivalent Fractions

Display a chocolate bar in its wrapper, identifying it as 1 whole bar. **This bar is made of smaller, bite-sized pieces of equal size. How many do you think there are?** (Allow students to guess. Open the bar and count the 12 pieces.)
What fractional part of the whole bar is 1 of the small pieces? ($\frac{1}{12}$) **How many are in the whole bar?** (12) **What fraction represents the whole bar?** ($\frac{12}{12}$) Write on the board: $1 = \frac{12}{12}$
If this bar were to be shared among friends without breaking any of the pieces into smaller pieces, could 2 friends share the bar? (yes) **How much would each person get?** ($\frac{1}{2}$ or $\frac{6}{12}$) Have each group make a model of a 12-piece candy bar by putting together 12 interlocking cubes and breaking the 12 cubes into two parts, showing that each part can be named as $\frac{1}{2}$ or $\frac{6}{12}$.
Ask students work together using the cubes, to find if the bar could be shared by 3–12 friends. Have them complete the following table.

No. of Friends	Each person gets
2	$\frac{1}{2}$ or $\frac{6}{12}$
3	$\frac{1}{3}$ or $\frac{4}{12}$
4	$\frac{1}{4}$ or $\frac{3}{12}$
5	no
6	$\frac{1}{6}$ or $\frac{2}{12}$
7	no
8	no
9	no
10	no
11	no
12	$\frac{1}{12}$

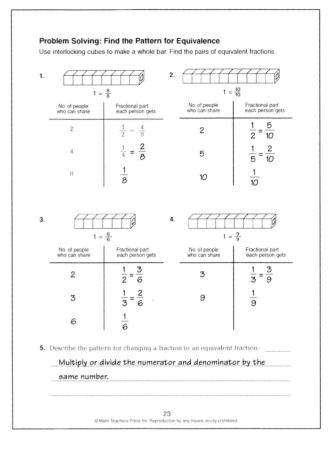

After the table is complete, ask how the amount of the bar represented by $\frac{12}{12}$ compares to the amount represented by $\frac{2}{2}$, how $\frac{1}{2}$ bar compares to $\frac{6}{12}$ bar, how $\frac{1}{4}$ bar compares to $\frac{3}{12}$, $\frac{1}{3}$ to $\frac{4}{12}$ and $\frac{1}{6}$ to $\frac{2}{12}$. (Each pair names the same amount.)
We call fractions with the same value but different names *equivalent fractions*. Can you find the pattern for changing a fraction to an equivalent fraction? What might you do to change $\frac{1}{2}$ to the equivalent fraction $\frac{6}{12}$? Multiply both terms by 6.

$$\frac{1 \times 6}{2 \times 6} = \frac{6}{12}$$

About This Page

Have students use interlocking cubes with a partner or small group to complete the page. Point out that they will be asked to describe the special pattern for changing a fraction to an equivalent fraction in problem 5.

Journal Prompt

Which fraction does not belong in this sequence of equivalent fractions? Explain why this fraction does not belong. Replace it with the correct fraction.

$\frac{1}{3}, \frac{2}{6}, \frac{3}{9}, \frac{4}{16}, \frac{5}{15}$

Objective: To find the greatest common factor of two numbers. To change a fraction to lowest terms.

Materials: Interlocking cubes, square tiles (or squares cut from Master 7)

Vocabulary: factor, common factor, simplify, greatest common factor, Venn diagram

Introductory Activities

Finding the Greatest Common Factor

Distribute interlocking cubes or square tiles to each pair. Find all the factors of a number by building arrays.

To find the factors of 8, take 8 cubes or tiles and arrange them in as many rectangles as possible. (1 × 8 or 2 × 4) **The sides of the array are the factors of 8.**

Now find the factors of 12 by building all possible rectangles with 12 cubes and listing the sides of each rectangle. (1, 12, 2, 6, 3, 4)

Write on the chalkboard:
   Factors of 8   Factors of 12
   1, 8, 2, 4   1, 12, 2, 6, 3, 4

Which factors are common or are in both sets? (1, 2, 4)

Which is the greatest common factor? (4)

We can show this with a special drawing called a Venn diagram.

   Factors of 8   Factors of 12
   (1, 8, 2, 4)   (1, 12, 2, 6, 3, 4)

   Common Factors of 8 and 12
   (8 (1, 2, 4) 3, 6, 12)

The <u>Greatest</u> Common Factor is 4.

We can use the greatest common factor, 4, to change the fraction ⁸⁄₁₂ to lowest terms.
   8 ÷ 4 =   2
   12 ÷ 4 =   3

Example: Simplify ²⁰⁄₂₄.

Have students find the greatest common factor for 20 and 24 (4). Then ask a student to draw a Venn diagram to show the greatest common factor (4). Complete the problem by dividing both terms by 4.
$$\frac{20 \div 4}{24 \div 4} = \frac{5}{6}$$

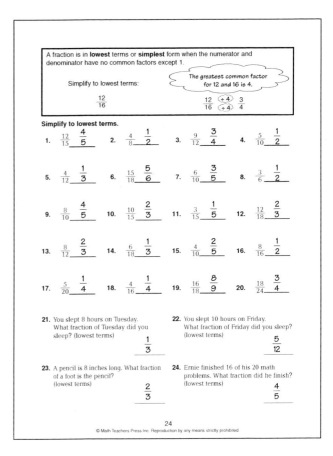

About This Page

Allow students to use interlocking cubes to identify the greatest common factor to be used to simplify each fraction.

Follow Up Activities

Divisibility Rules

Have students list the multiples of 2 from 2 to 30. (2, 4, 6, 8, 10, ... 28, 30) **These are even numbers. They can all be divided exactly by 2, with no remainder. We say they are "divisible by 2." Can you find the pattern that tells us when a number is divisible by 2?** (The numbers end in 0, 2, 4, 6, or 8.) Next ask students to find the patterns for the numbers 3, 4, 5, 6, and 8. (**3:** The sum of the digits is divisible by 3; e.g., for the number 462, 4 + 6 + 2 = 12 and 12 is divisible by 3.) (**4:** The number must be an even number and the last two digits divisible by 4.) (**5:** The numbers end in 5 or 0.) (**6:** The number must be an even number and the sum of the digits divisible by 3, e.g., 9555, because 9 + 5 + 5 + 5 = 24, and 24 is divisible by 3.) (**8:** The number must be an even number and the last three digits divisible by 8; e.g., 6712 is divisible by 8 because 712 is divisible by 8.)

Skill Builders 12-1, 12-2

Objective: To compare and order fractions with different denominators. To find the least common denominator.

Materials: Fraction Bars®, 10-sided dice

Vocabulary: least common denominator

Introductory Activities

Comparing Unlike Fractions Visually

Write on the board: Five friends each bought a large box of popcorn. After 15 minutes, the boxes were ¼, ½, ¾, ⅛ and ⅝ full. **Put the boxes in order from least to greatest.** Ask students to suggest various ways to solve the problem. Some may suggest drawing pictures, using fraction bars, measuring cups.

Arrange class into groups of two, each with a set of fraction bars.

Select all the bars with exactly one part shaded. Put them in order from least to greatest value. **Which is the least shaded?** (¹/₁₂) **On which side of your desk should you put ¹/₁₂?** (on the left) **Which fraction comes next?** (¹/₁₀)

After the bars have been ordered, have students write the names for the fractions in order: ¹/₁₂, ¹/₁₀, ⅙, ⅕, ¼, ⅓, ½.

Can you describe the how the value of a fraction changes when the top term or numerator stays the same and the bottom number decreases? (If the value of the denominator is decreasing and the numerator stays the same, the fraction is increasing.)

It is important for students to discover and verbalize this inverse relationship, because they tend to say ⅛ > ⅓ when they see it on a test.

Discover the Pattern: Equivalent Rates

Sometimes we compare two fractions and we do not have fraction bars or other models to visually compare them. Then we must look for a pattern to help us compare the fractions.

For example, if we know that one student makes ¾ of her free throws and another student makes ⁵/₇, how could we find who has the better free throw rate?

Write on the board: Which is greater, ¾ or ⁵/₇ ?

Have students work in small groups to decide which is greater and explain their decision. When students are allowed to make their own

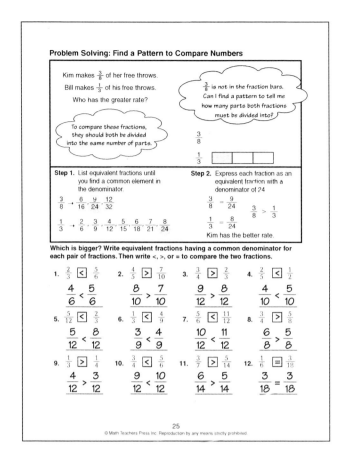

discoveries, they will often discover or reinvent the algorithm you wish to teach.

In this problem, students find a common element (denominator) by generating equivalent fractions for each fraction, until there is a common denominator for both sets.

¾ = ⁶/₈ , ⁹/₁₂ , ¹²/₁₆ , ¹⁵/₂₀ , ¹⁸/₂₄ , ²¹/₂₈
⁵/₇ = ¹⁰/₁₄ , ¹⁵/₂₁ , ²⁰/₂₈
²¹/₂₈ > ²⁰/₂₈

Assessment

What is one way you can use to tell if one fraction is greater than another? (Change each fraction to an equivalent fraction that has the same number of parts or the same denominator. Compare the numerators.)

About This Page

Students may use fraction bars to compare the fractions for the first 4 problems. To compare ⅔ to ⅚, students might place the ⅔ bar on the ⅚ bar and see that it is less. Or the yellow ⅔ bar may be changed to the red ⁴/₆ bar. ⁴/₆ < ⅚.

Follow Up Activities

Skill Builders 13-1

Objective: To estimate with fractions.

Materials: Fraction Bars®

Introductory Activities

Estimation with Fractions

On a recent National Assessment of Educational Progress (NAEP), only 24% of the 13-year-olds could correctly estimate the best answer for $^{12}/_{13} + ^{7}/_{8}$. Students need experiences seeing fractions that are close to 1 whole and fractions that are close to zero.

Have students sort a set of fraction bars into piles that are close to 1 whole ($^{11}/_{12}$, $^{9}/_{10}$, $^{5}/_{6}$, $^{4}/_{5}$) and those that are close to 0 ($^{1}/_{12}$, $^{1}/_{10}$, $^{1}/_{6}$, $^{1}/_{5}$).

Sometimes we do not need an exact answer. We need to know if the fraction is close to the number 1 or if it is just a small part of the number 1.

Write several addition and subtraction word problems on the board, asking students to estimate the answer.

Examples:
One can of peanuts has $^{3}/_{4}$ cups of peanuts. Another can has $^{7}/_{8}$ cups. About how many cups of peanuts do you have?
$^{3}/_{4} + ^{7}/_{8}$ = (about 1 + 1 or 2)

You spend $1^{1}/_{12}$ hours on homework on Monday and $2^{9}/_{10}$ hours on Tuesday. About how much time do you spend on both days?
$1^{1}/_{12} + 2^{9}/_{10}$ = (about 1 + 3 or 4)

You live $4^{1}/_{4}$ miles from the school. After you walk $1^{7}/_{8}$ miles, about how far are you from school?
$4^{1}/_{4} - 1^{7}/_{8}$ = (about 4 − 2 or 2)

About This Page

Direct attention to the example at the top of the page. Discuss the rounding of each mixed number to the nearest whole number.

Students should round each fraction to the nearest whole number.

Read several problems together, asking volunteers to round each fraction to the nearest whole number. Students may then complete the page on their own.

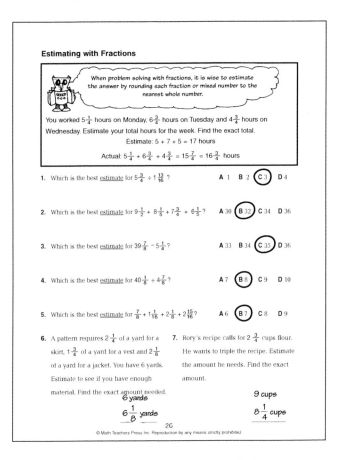

Follow Up Activities

Writing Word Problems

Ask students to make up two word problems requiring the use of estimation in addition and subtraction. Discuss and solve several of the problems. Save the problems for the class file.

Journal Prompt

Sort these fractions into three groups: close to 0, close to 1, more than 1:
$^{7}/_{8}$, $^{5}/_{6}$, $^{1}/_{10}$, $1^{1}/_{12}$, $^{1}/_{8}$, $^{10}/_{6}$, $^{9}/_{10}$, $^{1}/_{2}$

Justify why you placed each fraction into each category.

Objective: To add and subtract like fractions. To simplify proper fractions to lowest terms.

Materials: Tortilla rounds, Fraction Bars®, overhead Fraction Bars® (optional), plastic knives

Skill Builders Vocabulary: lowest terms

Introductory Activities

Problem Solving: Using Food as Models

Tortillas make appropriate models for demonstrating fraction problems. Give each small group one or more tortillas and a plastic knife.

Each tortilla represents 1 whole or 1 unit. How would it be shared with two people? (Cut it into two parts of equal size.)

How would you share it with four people? (Cut into four equal parts.)

Cut your tortilla to show how it would be shared with four people.

Suppose your tortilla were a pizza and you ate two of the four pieces, what fractional part of the whole pizza would you have eaten? ($2/4$ or $1/2$)

Suppose you eat one more piece, what fractional part of the whole pizza would you have eaten? ($3/4$)

Write the problem on the board ($2/4 + 1/4 = 3/4$) and ask, **What is the pattern for adding fractions with equal parts or same denominators?** (Add the numerators.)

Writing Problems

Have each group make up one addition problem and one subtraction problem about the tortilla. Share the problems with the class.

About This Page

Read the example at the top of the page. Use tortilla pieces to demonstrate addition of fractions in problem 1 and subtraction in problem 5.

Use fraction bars for problems 2 and 6. Note that when subtracting, one bar is placed directly under the other and the question is asked, **How many parts do not match?** The answer is the part that is left over or does not match.

For problem 8, ask students to change the answer to lowest terms.

Review the pattern for changing a fraction to a simple fraction in lowest terms. Remind students to always check their fractional answers to see if

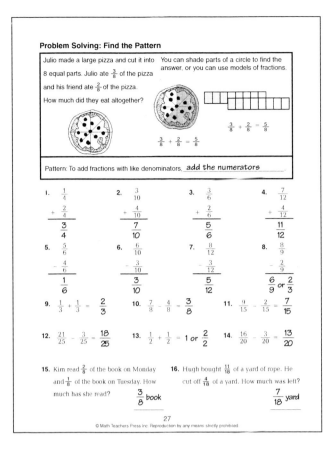

the numerator and denominator have a common factor.

Follow Up Activities

Make the Greatest Difference

Separate the class into groups of 2 or 4 players each with a set of fraction bars. Spread the bars face down. Each player takes turns selecting two bars of the same color and subtracting the two fractions. The player with the greatest difference after each round earns all the cards. If there is a tie, the bars are placed aside and the winner of the next round claims all the bars. When no additional rounds are possible, the player with the most bars wins the game.

Skill Builders 15-1

Objective: To add mixed numbers. To simplify in lowest terms.

Materials: Fraction Bars®

Introductory Activities

Addition and Subtraction of Mixed Numbers

Each time a fraction problem requires additional steps, e.g., changing from an improper fraction to a mixed number and then changing to lowest terms, the difficulty of the problem increases and the percentage of students able to answer the question correctly decreases. Remind students to always check their fractional answers to be sure they are in simplest form.

Problem Solving: Drawing Pictures

Students will develop understanding of fraction concepts and computations by drawing pictures of circles or rectangles to represent each number in a fraction problem. Make up a real problem that will require the addition of $1\frac{2}{4}$ and $2\frac{1}{4}$ in its solution. Have students read the problem aloud, find the questions and needed facts, and decide on the process.

A recipe calls for $1\frac{2}{4}$ cups of sugar and $2\frac{1}{4}$ cups of flour. How much sugar and flour are needed?

Students can draw pictures of shaded circles next to each addend and the sum.

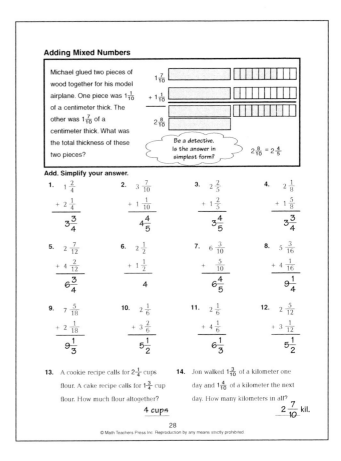

About This Page

Read and discuss the problem in the illustration at the top of page. Have students find answers to the first row of problems by drawing pictures.

Have volunteers show their drawings on the chalkboard.

Follow Up Activities

Drawing Pictures

Ask students to solve each problem by drawing a picture.

Skill Builders 18-1

Objective: To subtract mixed numbers. Includes renaming improper fractions and mixed numbers.

Materials: Paper plates, Fraction Bars®

Introductory Activities

Using Models, Drawing Pictures

Make up a problem related to $2\frac{3}{4} - 1\frac{2}{4}$.

Example: you have $3\frac{1}{4}$ pounds of butter. A recipe calls for $\frac{2}{4}$ cup of butter. How much will you have left?

Use small paper plates to represent $3\frac{1}{4}$. Have students draw a picture of $3\frac{1}{4}$. Ask how you could remove $\frac{2}{4}$ from $3\frac{1}{4}$. (Exchange 1 whole for $\frac{4}{4}$ and then remove $\frac{2}{4}$.) Have students draw a picture of the problem after the grouping:

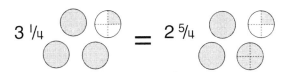

Then, cross out two of the fourths. How much is left? ($2\frac{3}{4}$). Show the recording on the board.

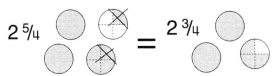

About This Page

Direct attention to the problem at the top of the page. Demonstrate how to solve the problem by drawing pictures and by using fraction bars or paper-plate circles. (One whole bar or circle must be exchanged for a $\frac{4}{4}$ bar or a circle before you can subtract.)

Demonstrate one or more problems with fraction bars.

Follow Up Activities

Writing Word Problems

Continue developing a class set of word problems by having students write at least one addition problem and one subtraction problem that might be solved by one of the computations on this page. Encourage students to include one- and two-step problems involving fractions in real-life situations, e.g., cooking, carpentry, map distances, or the stock market. Problems with hidden facts and unnecessary facts should also be included.

Skill Builders 16-1, 16-2

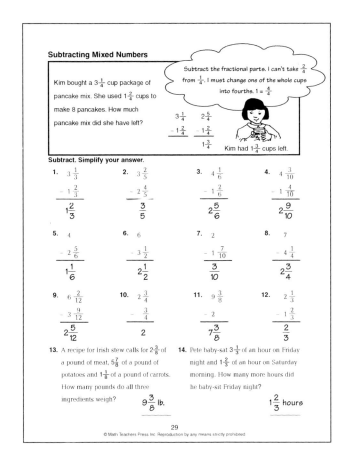

Objective: To add fractions with unlike denominators. DVD

Materials: Fraction Bars®, 10-sided dice, yellow and white interlocking cubes

Vocabulary: unlike fractions

Introductory Activities

Adding Unlike Fractions

Make up a problem related to $\frac{1}{3} + \frac{1}{2}$.

Example: Tish ate $\frac{1}{3}$ of a whole pizza. Jake ate $\frac{1}{2}$ of the same pizza. How much of the pizza did they eat?

Ask students to work in small groups with a set of fraction bars to find the answer. Guide discussion of their solution to The Golden Rule of Fractions: To add or subtract fractions with different denominators, always change the fraction bars to the same color.

Have students display $\frac{1}{3}$ (yellow) and $\frac{1}{2}$ (green). **Can we add a yellow bar and a green bar?** (No, they must be the same color.) **Can you change $\frac{1}{3}$ and $\frac{1}{2}$ to the same color bar? What color?** (Change the bars to the red $\frac{2}{6}$ and red $\frac{3}{6}$). **How much is $\frac{3}{6} + \frac{2}{6}$?** ($\frac{5}{6}$)

Write the problem on the board:
Record:
$\frac{1}{3}$ (yellow) $\frac{2}{6}$ (red)
+ $\frac{1}{2}$ (green)   + $\frac{3}{6}$ (red)
                            $\frac{5}{6}$

About This Page

Read and discuss the illustration at the top of the page. Demonstrate how to solve the problem with fraction bars and the "Golden Rule of Fractions:" You can't add or subtract fractions unless they are the same color (same denominator). Have students work problems 1–4 with fraction bars.

Follow Up Activities

Dicey Differences

Players take turns throwing two 10-sided dice twice and forming a fraction using the smaller number for the numerator and the larger number for the denominator. The player with the greater difference earns one point. For example a player throwing a 1 and a 6 on the first throw and a 2 and a 3 on the second throw would subtract: $\frac{2}{3} - \frac{1}{6}$ for a difference of $\frac{1}{2}$.

Ratio and Proportion in Lemonade

You are making some lemonade from concentrate. You use 3 cans of water with 1 can of frozen concentrate. How many cans of water would you need with 2 cans of frozen concentrate? 3 cans? 4 cans? Use the yellow and white cubes to demonstrate.

yellow   1   2   3   4
white   3   6   9   12

The ratio of lemonade to water is 1 to 3. All the recipes are proportional.

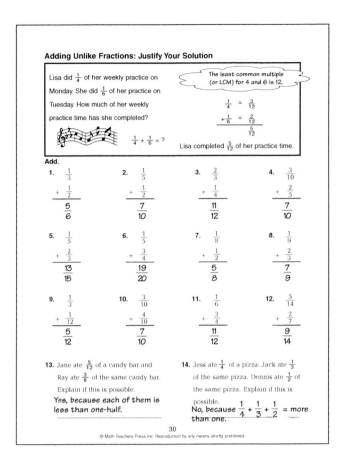

Objective: To subtract fractions with unlike denominators.

Materials: Multiple strips, Inch Graph Paper (Master 7), Fraction Bars®

Introductory Activities

Subtracting Unlike Fractions

Have students use 2 copies of Inch Graph Paper (Master 7) to make tables of the multiplication facts (also called tables of multiples).

×	1	2	3	4	5	6	7	8	9
1	1	2	3	4	5	6	7	8	9
2	2	4	6	8	10	12	14	16	18
3	3	6	9	12	15	18	21	24	27
4	4	8	12	16	20	24	28	32	36
5	5	10	15	20	25	30	35	40	45
6	6	12	18	24	30	36	42	48	54
7	7	14	21	28	35	42	49	56	63
8	8	16	24	32	40	48	56	64	72
9	9	18	27	36	45	54	63	72	81
10	10	20	30	40	50	60	70	80	90
11	11	22	33	44	55	66	77	88	99
12	12	24	36	48	60	72	84	96	108

Fractional Equivalencies

Cut the multiplication table into 12 strips of multiples. Use the strips to find the lowest common denominator and equivalent fractions for each pair of fractions. To add $2/3 + 1/4$, place the 2 multiple strip over the 3 multiple strip and the 1 multiple strip over the 4 multiple strip.

| 2 | 2 | 4 | 6 | **8** | 10 | 12 | 14 | 16 | 18 | 20 | 22 | 24 |
| 3 | 3 | 6 | 9 | **12**| 15 | 18 | 21 | 24 | 27 | 30 | 33 | 36 |

| 1 | 1 | 2 | **3** | 4 | 5 | 6 | 7 | 8 | 9 | 10 | 11 | 12 |
| 4 | 4 | 8 | **12** | 16 | 20 | 24 | 28 | 32 | 36 | 40 | 44 | 48 |

Your strips now tell you many equivalent fractions for $2/3$ and $1/4$. What are some equivalent fractions for $2/3$? ($4/6$, $6/9$, $8/12$, $10/15$, etc.) **What are some equivalent fractions for $1/4$?** ($2/8$, $3/12$, $4/16$, $5/20$, etc.)

When you are adding or subtracting fractions, they must always be divided into the same number of parts.

Fraction bars of the same color have been divided into the same number of parts. When writing fractions, this least number of parts is actually the denominator of the fraction and is sometimes called the lowest common denominator.

What is the lowest common denominator in our problem $2/3 + 1/4$? (The answer is 12 because it is the first denominator that is the

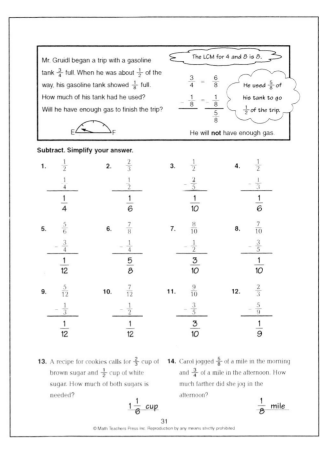

same using the multiple strips.) **What is the problem and answer?** ($8/12 + 3/12 = 11/12$)

Have students work in pairs. Have both students pick any two numbers from 2 to 12 and compare the two strips of multiples. Have them find the least common multiple for each number pair.

About This Page

Read the problem together. Have students use multiple strips to subtract the problems. Problems 1–4 may be worked with fraction bars and then verified with multiple strips.

Follow Up Activities

Journal Prompt

Write at least three statements to explain what to do when the denominators are unlike and you need to add or subtract fractions.

Skill Builders 17-1, 17-2

Objective: To find the pattern for multiplying fractions. To relate "of" to the operation of multiplication.

Materials: Paper, crayons

Vocabulary: product, of

Introductory Activities

Discover the Pattern: Multiplying Fractions

Students can learn the algorithm or rule for multiplying fractions with little conceptual understanding, because the ideas they learned for whole number multiplication also work with fractions (multiply the top numbers; multiply the bottom numbers). Therefore, instruction should emphasize modeling, finding a pattern and explaining why a fraction of a fraction or a fraction of a whole number will be less than the original number.

Paper Folding

Write on the board: You want to make $\frac{1}{2}$ of a recipe. How much of each ingredient do you use?
  Chocolate Chip Cookies
    2 eggs
    1½ cup flour
    ¼ cup oil
    ⅔ cup brown sugar
    ¾ cup oatmeal
    ⅔ cup sugar
    1 cup chocolate chips
    ½ cup raisin
    ½ teaspoon baking soda
    1 teaspoon vanilla
  Makes 60 cookies.

Give each student a sheet of paper. **Pretend that this paper represents 1 measuring cup. Could you fold your paper to show ½ cup of raisins? Now fold your paper to show the part that would be needed to make ½ of a recipe. What fractional part of your paper shows ½ of ½?** (¼) Write on the chalkboard: ½ of ½ = ¼ cup of raisins.
  Repeat, asking how much flour, oil, brown sugar and oatmeal would be used in ½ of the recipe (½ of ¼ = ⅛ and ½ of ⅓ = ⅙). Each time write the number sentence on the chalkboard and substitute the multiplication sign for the word "of."

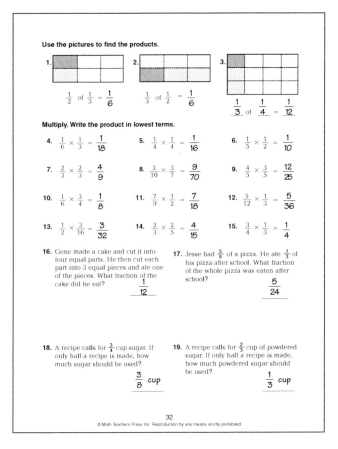

Why does it make sense to say that "of" means multiplication? (Because multiplying the denominators gives you the total number of parts and multiplying the numerator gives you the number of these parts taken.) **What is the pattern for multiplying fractions?** (Multiply the numerators and multiply the denominators. Simplify if possible.)

About This Page

Use paper folding to solve problems 1–3. For problem 1, have students fold a paper into thirds and label each part ⅓. Color ⅓ of the paper. Have students then fold the thirds into 2 equal parts, open the paper, and draw lines on the part which is ½ of ⅓. What is the fraction name for ½ of ⅓? (⅙) Repeat with problems 2 and 3.

Follow Up Activities

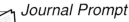 *Journal Prompt*
When two fractions are multiplied, what can you say about the size of the resulting fraction? Give an example to justify what happens. Use diagrams, words and/or symbols.

Skill Builders 19-1

Objective: To find a fractional part of a whole number.

Materials: Egg carton, pencils, cubes or eggs

Vocabulary: dozen, discount

Introductory Activities

Problem Solving: Acting out the Problem

Today we are going to find a way in which fractions are used to describe part of a set. What is the meaning of the word set? (a collection or a group of objects)

Have 12 students stand in front of the class. Ask how many students would leave if you were to ask one-half of them to go to the library. (6) Ask a student to explain how he got the answer. (I divided 12 into 2 groups of equal size and there were six in each group, or I thought about having 1 out of every 2 groups "of" students go to the library.)

Repeat with ⅓ of 12, ¼ of 12, ⅙ of 12.

Using a Model

Display a closed egg carton.

How many eggs are inside the carton when you buy it at the grocery store? (12 or 1 dozen) Open the carton and show 12 eggs or 12 cubes inside.

Sometimes grocery stores sell ½-dozen cartons so you can buy fewer eggs. How many eggs are inside a ½-dozen container? (6)

How did you get your answer?

After allowing students to discuss their answers, review the various methods shown at the top of the student page. In Method 1, show the open egg carton with a pencil dividing it into two equal parts. Then have students draw a picture of the eggs and pencil.

Remind students in Method 3 that a whole number is written as a fraction by writing the number over 1.

Point out that Method 4 uses a series of ratios to find that 1 out of 2 eggs is the same as 2 out of 4 eggs, is the same as 6 out of 12 eggs.

How many eggs are in ⅓ of a dozen? Use all four methods to find the answer.

How many eggs are in ⅔ of a dozen? Use all four methods to find the answer. Students

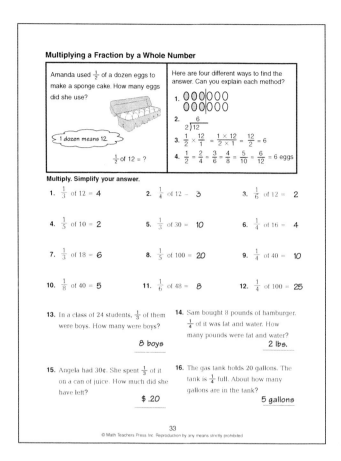

will note that to use Method 2, they must divide 12 by 3 and multiply the quotient 4 by 2. Use two pencils to divide the carton into thirds. Then place cubes or eggs in two of the one-third parts.

About This Page

After reading and solving the example at the top of the page, ask students how they think the answer to these two problems will compare: ½ of 12 and 12 of ½ (they will be the same). Ask a student to justify both answers by drawing a picture to show that the order property applies to the multiplication of fractions: ½ × 12 = 12 × ½.

Follow Up Activities

Skill Builders 19-2

6E Teacher Manual

Objective: To divide proper fractions, using models. To discover the pattern for division.

Materials: Fraction Bars®, 10-sided dice
Skill Builders Vocabulary: reciprocal

Introductory Activities

Division of Fraction Bars

Review the meaning of division with whole numbers to division with fractions.

In the division problem, $6 \div 3 = r$, what do the numerals 6 and 3 mean? (How many groups of 3 are in 6?)

What do you think $\frac{1}{2} \div \frac{1}{6}$ means? (How many sets of $\frac{1}{6}$s there are in $\frac{1}{2}$?) **Use your fraction bars to decide how many $\frac{1}{6}$s there are in $\frac{1}{2}$.** (3)

Write on the board: $\frac{1}{2} \div \frac{1}{6} = 3$

Repeat the questioning for several more division problems, having students use fraction bars to find the answers.

Ask students to look for a pattern to divide fractions. Appropriate examples: $\frac{1}{3} \div \frac{1}{6}$, $\frac{1}{2} \div \frac{1}{12}$, $\frac{2}{3} \div \frac{1}{12}$, $1 \div \frac{1}{6}$, $\frac{5}{12} \div \frac{1}{6}$.

The goal is to guide students to generalize two patterns for dividing fractions:
1. Change the bars to common colors; divide the numerators.
2. Invert the divisor and multiply. You may wish to share the following rhyme to help them remember the pattern: The number you divide by, turn upside down and multiply.

About This Page

Read the illustration together at the top of the page. Have students use fraction bars to solve the problems. Encourage students to estimate the answer before finding the exact answer.

Follow Up Activities

Dicey Fraction War

Separate the class into groups of two, and give each group a pair of 10-sided dice. Players take turns throwing the pair of dice twice, each time forming a fraction with the smaller number in the numerator and the larger number in the denominator. The die is thrown again each time a

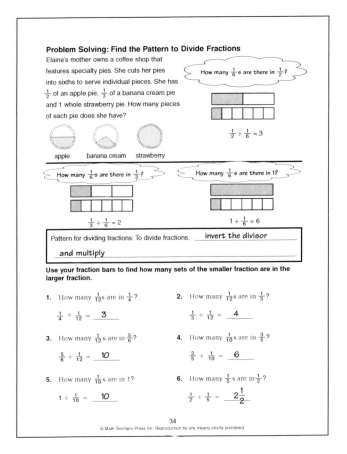

zero is shown. Using these two fractions, players write a division of fractions problem and solve. The player with the greater quotient is the winner.

Example: Player A throws 5 and 8 and writes the fraction $\frac{5}{8}$ on paper.

Player A throws 2 and 9 and writes the fraction $\frac{2}{9}$ on paper. He may write the number sentence $\frac{5}{8} \div \frac{2}{9} =$ or $\frac{2}{9} \div \frac{5}{8} =$

(They discover that writing the greater fraction first yields the greater quotient, so they choose to write $\frac{5}{8} \div \frac{2}{9} =$. Solution: $\frac{5}{8} \times \frac{9}{2} = \frac{45}{16} = 2\frac{13}{16}$.)

Player B throws 3 and 7 and writes $\frac{3}{7}$; Player B throws 1 and 4 and writes $\frac{1}{4}$; writes the number sentence $\frac{3}{7} \div \frac{1}{4}$ and solves, $1\frac{5}{7}$.

Player A receives 1 point for having the greater quotient.

In the beginning, you may wish to have groups of three for this activity with the third player serving as a checker, to see that the division is performed correctly. The first player to earn 5 points is the winner.

Skill Builders 20-1

Objective: To relate common fractions to decimal fractions. To write decimal fractions from number line models and vice versa.

Materials: Fraction Bars®
Vocabulary: decimal fraction

Introductory Activities

Equivalent Fractions and Decimals

Sort out the bars divided into tenths. Order your bars from least to greatest shaded. Write the name for each bar as a fraction and as a decimal. ($^0/_{10}$ or 0.0, $^1/_{10}$ or 0.1, $^2/_{10}$ or 0.2, $^3/_{10}$ or 0.3, $^4/_{10}$ or 0.4, $^5/_{10}$ or 0.5, $^6/_{10}$ or 0.6, $^7/_{10}$ or 0.7, $^8/_{10}$ or 0.8, $^9/_{10}$ or 0.9, $^{10}/_{10}$ or 1.0)

Count aloud with me by tenths from 0 to 1. ($^0/_{10}$, $^1/_{10}$, ... $^{10}/_{10}$ or 1)

Use your fraction bar to draw a picture of a number line from 0 to 1 divided into tenths. Write a fraction above and a decimal fraction below each mark on your number line.

Demonstrate on the chalkboard by drawing a long line and marking off a unit of 1 fraction bar on the line. Then, place a fraction bar on top of the line and mark each $^1/_{10}$ space from 0 to 10.

Sort out the bars divided into fifths. Order your bars from least to greatest shaded. Write the name for each fraction bar as a fraction. ($^0/_5$, $^1/_5$, $^2/_5$, $^3/_5$, $^4/_5$, $^5/_5$) Count aloud with me by fifths from 0 to 1. (zero-fifths, one-fifth, two-fifths, ... five-fifths or one.)

Can the fraction bars divided into fifths be written as decimals? (Yes, if they are changed to equivalent fraction in tenths).

How would you write $^1/_5$ as a decimal? (0.2) Show on the overhead that $^1/_5$ is equivalent to $^2/_{10}$ by comparing the shaded amounts of both fraction bars.

How would you write $^0/_5$, $^2/_5$, $^3/_5$, $^4/_5$, $^5/_5$ as decimals? (0.0, 0.4, 0.6, 0.8, 1.0)

Write the fractions from $^0/_5$ to $^5/_5$ on your number line above their equivalent fractions in tenths.

The completed decimal number line should look like this.

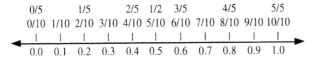

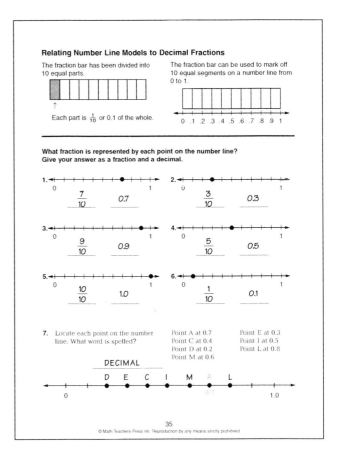

Are there any other fraction bars in your set which can be renamed as decimal fractions? (equivalent fraction bars for 0.0: $^0/_2$, $^0/_3$, $^0/_4$, $^0/_5$, $^0/_8$, $^0/_{12}$; equivalent fraction bars for 0.5: $^1/_2$, $^2/_4$, $^3/_6$, $^6/_{12}$; equivalent fraction bars for 1.0: $^2/_2$, $^3/_3$, $^4/_4$, $^6/_6$, $^{12}/_{12}$)

About This Page

Read the example at the top of the page. Emphasize that, when naming a point on a number line, the spaces between the marks on the line—not the marks—are counted. In problem 1, ask how many spaces between 0 and 1. (10) Each space is equal to $^1/_{10}$ or 0.1. What fraction and decimal fraction is shown by the point on the number line? ($^7/_{10}$ or 0.7)

Follow Up Activities

Skill Builders 21-1, 22-1

Objective: To identify the place value of a digit in the tenths or hundredths place. **DVD**

Materials: Base ten blocks, Place Value Mats (Masters 10 and 11), Coins and Bills (Master 8)

Introductory Activities

Money and Decimal Place Value

Display a $1 bill. **How much money is shown?** (one dollar) **How do you write the amount of money with a $ sign?** ($1.00) The one-dollar bill is the unit or whole or one.

Display a dime. **How much money is shown?** (10¢) **How many dimes does it take to make a dollar?** (10) **What fractional part of a dollar is a dime?** (one-tenth) **How do you write that amount of money with a decimal point?** ($0.10)

Display a penny. **How much money is shown?** (1¢) **How many pennies are in one dollar?** (100) **What fractional part of a dollar is one penny or one cent?** (one hundredth) **How do you write that amount of money with a decimal point?** ($0.01)

Display a $1 bill, 3 dimes and 2 pennies. **How much money is shown?** (one dollar and thirty-two cents) **How do you write this amount of money with a cents sign and a dollar sign?** (132¢ or $1.32)

Notice that the digit in the one dollar place shows the number of ones. The 3 shows the number of dimes or tenths of a dollar. The 2 shows the number of pennies or hundredths of a dollar.

Write each place value name under each digit to introduce the first two decimal place values.

$1 . 3 2
ones dimes pennies
 tenths hundredths

A Place Value Mat for Decimals

Have students fold a sheet of paper into three parts to make three columns for a place value chart. **Label each column as "Units or Ones," "Tenths" and "Hundredths." Draw a decimal point between the units and tenths.**

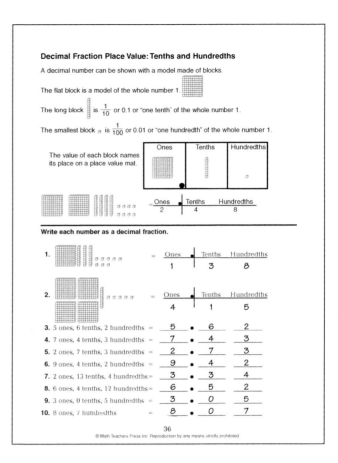

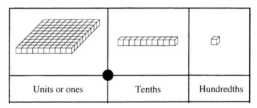

Write on the board: 1.34.

Have students use base ten blocks to build the number on the mat. **What digit is in the ones place?** (1) **What digit is in the tenths place?** (3) **What digit is in the hundredths place?** (4) Repeat with other numbers.

A dot may be glued or taped on the place value chart. Emphasize that everything to the left of the decimal point will be whole numbers and that everything to the right relates to decimal fractions less than one whole.

Build decimal mixed numbers with base ten blocks. Have students write the number to match each block.

Example: 1 flat, 2 rods (or long blocks), 6 units = 1.26

About This Page

Use base ten blocks to demonstrate the examples at the top of the page. Have students use base ten blocks to complete the page.

Objective: To identify the place value of a digit in the thousandths and ten-thousandths place.

Materials: Base ten blocks, paper, 1-centimeter squares cut from graph paper (Master 2), index cards, Place Value Bingo (Master 1), Bingo markers (corn or beans)

Vocabulary: thousandths place, ten-thousandths place

Introductory Activities

Connections: Concrete to Abstract

Have students fold a sheet of paper into four parts to make four columns for a decimal place value mat. Label each column as "Units or Ones," "Tenths," "Hundredths," and "Thousandths." Draw a large decimal point between the ones and the tenths or use a self-adhesive dot.

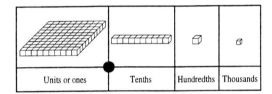

Write on the board: 2.437

Have students build the number on the mat. Use flats for the ones, longs for the tenths, unit blocks for the hundredths and centimeter squares cut from Master 2 or small squares cut from needlepoint canvas for the thousandths. **What digit is in the ones place?** (2) **What digit is in the tenths place?** (4) **What digit is in the hundredths place?** (3) **What digit is in the thousandths place?** (7) Repeat with other decimals, e.g., 1.054, etc.

About This Page

Read the examples together at the top of the page. Use blocks and centimeter graph paper (Master 2) to illustrate each of the place values in the place value picture.

Follow Up Activities

Place Value Bingo

Refer to Card 4 of Master 1 for instructions for this game. Prepare 27 index cards by writing the following on each card: 1 tenths, 2 tenths, ... 9 tenths;

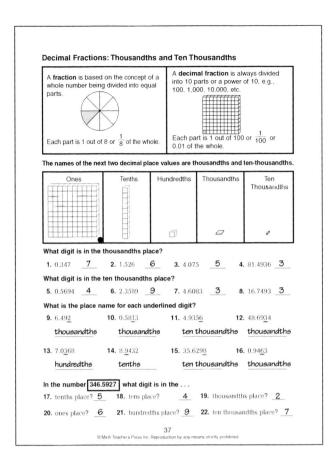

1 hundredth, 2 hundredths, ... 9 hundredths; 1 thousandth, 2 thousandths, ... 9 thousandths.

Have each student prepare a Place Value Bingo card by drawing a tic-tac-toe grid, labeling the column from left to right as tenths, hundredths, thousandths, and writing one of each of the digits 1–9 randomly in the nine squares of the grid.

Tenths	Hundredths	Thousandths
3	5	2
7	6	8
1	9	4

Shuffle the Bingo cards. Select and read aloud one card at a time, pausing to allow students time to cover the digit with a marker if it is in the right place. With the above Bingo card, when the teacher reads "6 tenths," the student cannot cover the 6 because it is in the hundredths place, not the tenths place.

The first player to cover three in a horizontal row calls out, "Bingo!" reads his number, e.g., three hundred fifty-two thousandths, and is the winner of that round. Play several rounds.

Skill Builders 23-1, 23-2

Objective: To order and compare decimal fractions of the same length.

Materials: Overhead money (optional), base ten blocks, string, Fraction Bars®, coins and bills (or Master 8)

Introductory Activities

Using Models

Make 2 string circles. Display 3 dimes inside one circle and 7 dimes inside the other. **How much money is shown inside each circle? How do 3 dimes compare to 7 dimes?** (3 dimes is less than 7 dimes.)
Write on the board: .30 < .70
Display a $1 bill, 3 dimes and 7 pennies inside one string circle and a $1 bill, 4 dimes and 2 pennies inside the other. **How much money is shown inside each circle?** ($1.37 and $1.42) **How does $1.37 compare to $1.42?**
Write on the board: 1.37 < 1.42

Comparing and Ordering Fraction Bars

Select the set of fraction bars divided into tenths. Order the fraction bars from least to greatest shaded across your desk. Count aloud with me by tenths from 0 to 1. (⁰/10, ¹/10, ²/10, ³/10, ⁴/10, ⁵/10, ⁶/10, ⁷/10, ⁸/10, ⁹/10, ¹⁰/10 or 1)

Comparing and Ordering Base Ten Blocks

Build 4 tenths on your desk. Line the blocks up end to end. Now build 2 tenths under them lining the blocks up end to end. Now build 5 tenths under them, lining the blocks up end to end.

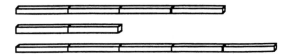

Which is the least? (2 tenths) **next least?** (4 tenths) **greatest?** (5 tenths) **Write the decimals in order from least to greatest.** (0.2, 0.4, 0.5)
Repeat with other groups of tenths.
Build 3 hundredths on your desk, lining the blocks up end to end. Build 7 hundredths, lining the blocks up end to end. Build 6 hundredths, lining the blocks up end to end.
Write the decimals in order from least to greatest. (0.03, 0.06, 0.07)
Repeat with other groups of hundredths.

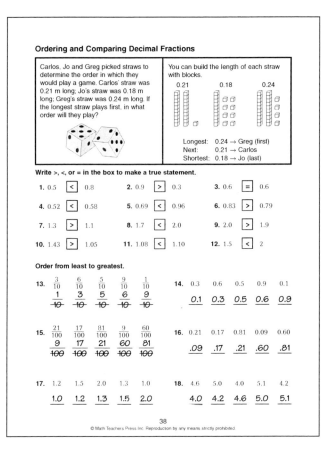

About This Page

Use base ten blocks to illustrate the length of each of the three "straws" in the illustration. A straw .21 m long is shown with 2 tenths blocks and 1 hundredth block. Ask which is shortest, next shortest and longest.

Follow Up Activities

Fraction War

Deal an equal number of white fraction bars face down to each player. Players turn over one bar at a time. The player with the greater part shaded wins them both by saying the correct inequality aloud, e.g., ⁷/10 is greater than ³/10.

Journal Prompt

Four students were playing a game called Decimal War. The first player drew a card that showed 5.1; the second player drew 3.5; the third player drew 3.1; and the fourth player drew 0.5. The winner of the round is the player who draws the greatest decimal. Who is the winner of this round? Explain why this player is the winner.

Skill Builders 24-1

Objective: To find the pattern for writing equivalent decimal fractions. To compare decimal fractions of uneven lengths.

Materials: Base ten blocks, 6-sided dice

Introductory Activities

Counting by Hundredths

Have students use base ten unit blocks to count by hundredths from 0.01 to 0.10. (Display 1 unit block and say "1 hundredth," then add one more unit block and say "2 hundredths." Continue this way until you reach 10 unit blocks and say "10 hundredths or 1 tenth.")

Repeat the pattern of counting by hundredths from 0.11 (1 tenths block with 1 hundredths block) to 0.20 (2 tenths blocks) and from 0.21 to 0.30.

Discover the Pattern: Equivalent Decimals

Build 1 tenth. Now build 10 hundredths under it.

How does the value of 1 tenth compare to 10 hundredths? (Both have the same value.)

Write on the board: 0.1 = 0.10. **One tenth is equivalent to 10 hundredths.**

Repeat this activity for 2 tenths and 20 hundredths, 3 tenths and 30 hundredths to show equivalent decimal fractions.

Ask students to explain how many times they exchanged 10 of the smallest blocks (hundredths) for 1 of the next larger block (1 tenth). Record equivalent fractions in fraction and decimal form for each of the exchanges:

fraction equivalent	decimal equivalent
1/10 = 10/100	0.1 = 0.10
2/10 = 20/100	0.2 = 0.20
3/10 = 30/100	0.3 = 0.30
"	"
"	"
10/10 = 100/100	1.0 = 1.00

What is the pattern for changing a fraction to an equivalent fraction? (Multiply or divide the numerator and denominator by the same number.)

Can you describe the pattern for writing a decimal equivalent to a given decimal? (Write zeros at the far right of the decimal number or remove zeros from the far right of any decimal numeral.) **How do you change a**

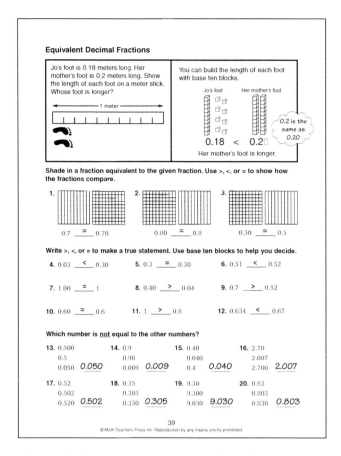

decimal fraction to an equivalent decimal? (Write zeros to the right of the number.)

Comparing Decimals of Uneven Length

Write on the board: A bag of chocolate raisins weighs 0.3 kilograms. A box of chocolate pretzels weighs 0.23 kilograms. Which weighs more? Ask students to discuss the problem with a partner and be prepared to tell which weighs more and how they know. Students may build 0.3 as 3 tenths and 0.23 as 2 tenths 3 hundredths. They will then see that 3 tenths is actually more than 2 tenths 3 hundredths, so 0.3 > 0.23.

About This Page

Read the information at the top of the page. Use base ten blocks to demonstrate 0.28 is less than 0.3.

Work Problems 1, 4 and 13 with the class. Have students complete the page on their own.

Follow Up Activities

Journal Prompt

Explain how to change a decimal fraction to an equivalent decimal. Give at least two examples to show that the rule works.

Objective: To interchange decimals and fractions with denominators of 10 or 100.

Materials: Coins and bills (or Master 8), Fraction Bars®, base ten blocks, Place Value Mats (Masters 10 and 11)

Introductory Activities

Equivalent Decimals and Fractions

Demonstrate how dimes and pennies name the first two decimal place values (tenths and hundredths). Display a $1 bill, 2 dimes and 5 pennies. **Write the dollar amount shown.** ($1.25) **What fractional part of a dollar is 1 dime?** ($1/10$) **We can call the dimes place the tenths place.** Write "tenths" under the 2. **What fractional part of a dollar is 1 penny?** ($1/100$) **We can call the pennies place the hundredths place.** Write "hundredths" under the 5.

Connect these ideas together by writing this chart on the board.

$1 bill	dimes	pennies
ones	tenths	hundredths
1s	$1/10$s	$1/100$s

Each student will need a decimal place value mat.
Build 2 tenths on your mat.

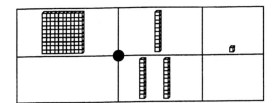

Write 2 tenths as a fraction and as a decimal. ($2/10$, 0.2) Write on the board: $2/10 = 0.2$
Build 8 hundredths on your mat.
Write the number as a fraction and decimal. ($8/100 = 0.08$)

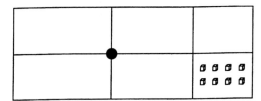

Say decimal fractions from $1/10$ to $9/10$ out loud, e.g., seven tenths.

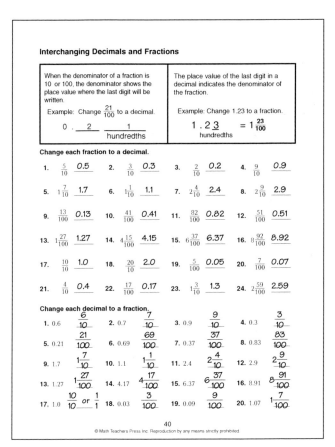

Have students build or draw what they hear. Have students write what they hear as a decimal and as a common fraction.

Repeat by saying decimals in hundredths, e.g., twenty-three hundredths, etc.

About This Page

Read the illustration together at the top of the page. Ask students to read each fraction or decimal fraction aloud as you write the numeral 0.21.

Direct attention to problem 1. Show $5/10$ with fraction bars and base ten blocks.

Follow Up Activities

Skill Builders 25-1

Objective: To add decimals in tenths. **DVD**

Materials: Base ten blocks and Place Value Mats (Masters 10 and 11)

Introductory Activities

Problem Solving: Using Models.

Write on the board: You walked ³⁄₁₀ of a kilometer to a friend's house. Then you and your friend walked ⁵⁄₁₀ of a kilometer to the theater. How far did you walk? **Build 3 tenths on the mat. Add 5 tenths. How many in all?** (8 tenths)
Write on the board:

$$\begin{array}{r} 0.3 \\ + 0.5 \\ \hline 0.8 \text{ km} \end{array}$$

Write on the board: You walked ⁴⁄₁₀ mile and then ⁸⁄₁₀ mile more. How far did you walk? **Build 4 tenths on your mat. Add 8 tenths. How many in all?** (12 tenths or 1 whole and 2 tenths)
Write on the board:

$$\begin{array}{r} 0.4 \\ + 0.8 \\ \hline 1.2 \text{ mile} \end{array}$$

Watch for students who incorrectly add 0.4 to 0.8 and write 0.12. **What is the sum of 4 tenths and 8 tenths?** (12 tenths) **Can you exchange or regroup?** (Yes, exchange 10 tenths for 1 whole.)

About This Page

Read the problem at the top of the page. Demonstrate the solution with base ten blocks, emphasizing that 10 of any 1 block must be exchanged for 1 of the next larger block before the answer is recorded. Demonstrate the shading to complete problem 1. Complete the first row of the addition table in Problem 3 with the class. Have students complete the page on their own.

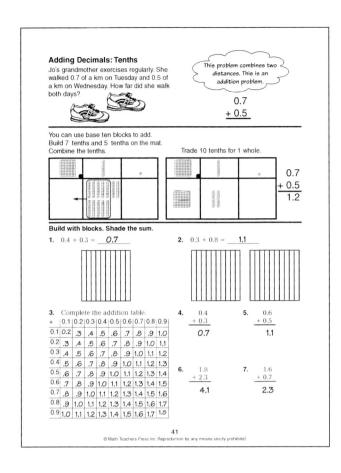

6E Teacher Manual   41

Objective: To add and subtract decimals in hundredths. **DVD**

Materials: Base ten blocks, coins and bills (or Master 8), overhead coins and bills (optional), decimal place value mats

Introductory Activities

Addition and Subtraction of Decimals

Read or write this problem: You live 1.3 miles from school. You ride your bike 0.5 mile and stop to rest. How far are you from school? **Build 1 whole and 3 tenths on your mat. Remove 5 tenths. How much is left?** (1 whole 3 tenths must be exchanged for 13 tenths, so that 5 tenths may be removed.)
Record

```
  1.3
- 0.5
  0.8 mile
```

Write on the board: A large cinnamon roll weighs 0.32 kilograms. A package of rolls weighs 0.45 kilograms. How much do they weigh together? 0.32 + 0.45

Have students build each decimal (3 tenths 2 hundredths and 4 tenths 5 hundredths) on a decimal place value mat and then "add" by

Units or ones	Tenths	Hundredths

combining blocks with the same value. Have a student record the problem and its sum in vertical format on the chalkboard. (7 tenths 7 hundredths, which is read 77 hundredths.)
How do you add or subtract decimals? (Combine or separate blocks of the same value, e.g., tenths with tenths and hundredths with hundredths.)
In addition of decimals, how do you know when that sum needs to be renamed? (When there are 10 or more of the same size block.) **In subtraction of decimals, when will you have to rename one of the numbers?** (When there are not enough of one block to remove from the original number.)

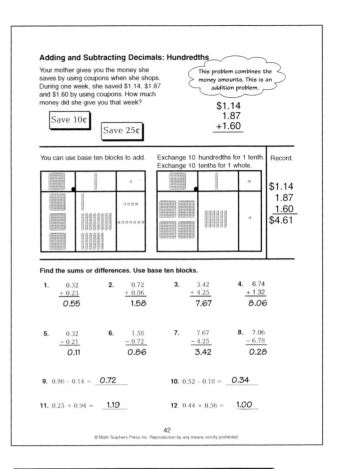

About This Page

Read the problem at the top of the page. Use overhead dollar bills, dimes and pennies to find the sum. Use base ten blocks to solve the problem. Then study the picture to see how the regrouping is shown. Have students use base ten blocks to solve the first 3 problems and complete the rest of the page with paper and pencil.

Follow Up Activities

Have students make Magic Squares of decimal numbers adding to a given number such as 1. The same number may be used only once. Players earn one point for each horizontal, vertical or diagonal line adding to 1. A maximum of 8 points may be won with each magic square.

Ex.	.15	.55	.30
	.25	.10	.65
	.60	.35	.05

This player earns 7 points because there are 7 combinations that add to 1 (3 horizontal, 3 vertical and 1 diagonal).

Objective: To solve 1- and 2-step word problems involving reading tables.

Materials: Base ten blocks, local fast-food menu, order pads, calculator, coins and bills (or Master 8), Place Value Mats (Masters 10 and 11)

Introductory Activities

Acting at the Play Restaurant

Have students work in pairs with a copy of a menu from a local fast food restaurant and several sheets of order pads. Have 1 person order a lunch of 2–4 items. Another student, the order taker, will record the order, the amount of each item and the total bill. Then have the first student pay the cashier with larger denomination bills, so the cashier has to make change.

Example: If your order comes to $7.95, the orderer may give the cashier a $10 bill or two $5 bills. The cashier returns 1 nickel and two $1 bills.

Subtracting Across Zero

Making change from $5, $10 and $20 bills requires knowing how to subtract across zero. Use base ten blocks to demonstrate the regrouping in subtraction.

Write on the board: Alex has $4.00. He buys a chicken sandwich for $2.61. How much does he have left? After reading the problem and deciding on the process of subtraction, write on the board:

$4.00
− 2.61

Show me $4.00 with the fewest possible number of bills or coins. (four $1 bills) **How many pennies in $2.61 need to be taken away?** (1) **How can we take 1 penny away from 0 pennies?** (Exchange 1 $1 bill for 10 dimes and then exchange 1 of those dimes for 10 pennies.) **Now we can remove 1 penny, 6 dimes and 2 $1 bills. How much is left?** ($1 bill, 3 dimes, 9 pennies or $1.39) Show how the two regroupings in this problem are recorded as you work the problem out loud on the chalkboard.

We can use base ten blocks instead of money to work the problem 400 − 261. Build 400 on your mats. Show the exchanging you

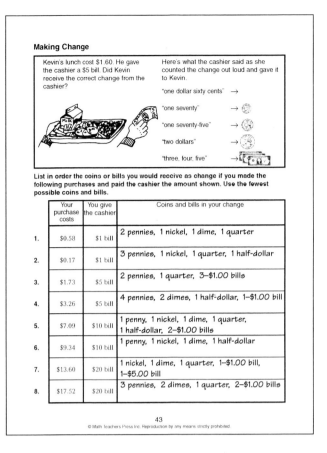

will need to do before you can remove 261. (Check the blocks on the mat) **Remove 261. What is left?** (139) Show how the two regroupings are recorded on the chalkboard.

About This Page

Read the problem at the top of the page. Use play coins and bills to show how to count out the change. Then show the problem with paper and pencil.

Follow Up Activities

Skill Builders 26-1, 43-1

Objective: To multiply a whole number by hundredths.

Materials: Base ten blocks and Place Value Mats (Masters 10 and 11), coins and bills (or Master 8)

Introductory Activities

Multiplying Whole Numbers and Decimals

Write on the board: Find the cost of 3 apples at 25¢ each. Have students use real or play money to build 3 groups of 25¢ with dimes and pennies on the Place Value Mat. **Put together the pennies and the dimes; exchange 10 pennies for 1 dime.** Ask a student to write the problem on the board and show the regrouping.

$$\begin{array}{r} \overset{1}{25} \\ \times\ 3 \\ \hline 75¢ \end{array}$$

Then show the solution of 3×0.25 with base ten blocks. Have a student write the problem in vertical format on the chalkboard and explain the regrouping of 10 of the 15 hundredths to 1 tenth in the recording of the product.

$$\begin{array}{r} \overset{1}{0.25} \\ \times\ \ 3 \\ \hline 0.75 \end{array}$$

About This Page

Read the problem in the illustration together. Use base ten blocks to demonstrate. Ask a student to describe the pattern for multiplying a whole number by hundredths. (Multiply as with whole numbers. Express the answer as hundredths, by moving the decimal point two places to the left.)

Follow Up Activities

Assessment

How do you multiply a decimal and a whole number? (Build the decimal the same number of times as the whole number; put together like blocks, regroup as necessary; multiply as with whole numbers. Have the same number of decimal points in the product as in the decimal.)

Skill Builders 27-1

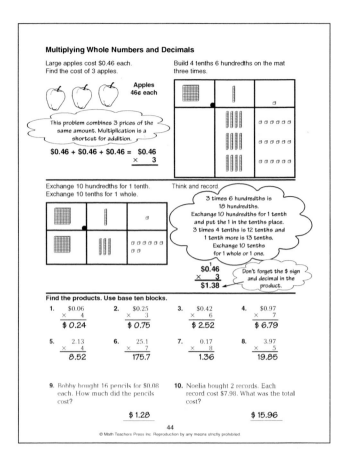

Objective: To generalize and apply a pattern for multiplying decimals.

Materials: Base ten blocks, centimeter squares (Master 2)

Introductory Activities

Multiplying Tenths by Tenths

Write on the board: John lives 0.1 kilometers from school. He walks 0.3 of the way and stops to buy a bagel at the corner. How far has he walked?

To help students visualize the problem, ask a student to act it out. **If the distance across our room is 0.1 or 1 tenth of a kilometer, how far would a person walk if they walked $^3/_{10}$ of the way?** The class should understand that where the student stops should be less than halfway, because 0.3 is less than 0.5. Then draw a picture of the problem on the chalkboard.

How can we find a point $^3/_{10}$ of the way between 0 and $^1/_{10}$? (Divide the line into 10 equal parts.) **What shall we name each part?** (0.00, 0.01, ... 0.10) **How much is 1 tenth of 1 tenth?** (1 hundredth) **How far has he walked?** (0.03 or 3 hundredths)

Use your base ten blocks to find 1 tenth of 1 tenth. (one hundredth) **Describe how you found your answer.** (I built 1 tenth. I had to exchange 1 tenth for 10 hundredths. Then I separated 1 out of the 10 hundredths.)

Multiplying Tenths by Hundredths

Cut one centimeter squares from copies of Master 2 and distribute to students to use as 1 thousandths blocks.

Write on the board:
0.1 × 0.01 = ?   0.3 × 0.01 = ?   0.3 × 0.02 = ?

Display 1 hundredth with base ten blocks. (1 unit block) **How much is 1 tenth of 1 hundredth?** (1 thousandth)

Show me 1 thousandth. (1 cm square) Say with me "1 tenth of 1 hundredth is 1 thousandth."

Record: 0.1 × 0.01 = 0.001 and $^1/_{10}$ × $^1/_{100}$ = $^1/_{1000}$.

Repeat with 0.3 × 0.01 and 0.3 × 0.02. **What is the pattern for multiplying tenths (a 1-place decimal) times hundredths**

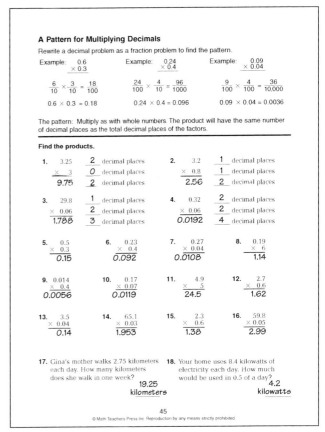

(a 2-place decimal)? (The product will be in thousandths, a 3-place decimal.)

Multiplying Decimals

0.3 × 2 =   0.3 × 0.2 =   0.3 × 0.02 =

Use base ten blocks to solve each problem. Record each answer. (0.6, 0.06, 0.006) **What is the pattern for the decimal point in the product?** (The number of digits to the right of the decimal point in the product will be equal to the sum of the number of digits to the right of the decimal point in the two numbers being multiplied.)

About This Page

Use the example at the top to review the pattern for multiplying tenths by tenths, tenths by hundredths and hundredths by hundredths.

Follow Up Activities

Journal Prompt

Explain the pattern for placing the decimal point in the product of two decimal numbers. Give two examples that show the pattern.

Skill Builders 27-2

Objective: To find the pattern to divide decimal fractions by whole numbers.

Materials: Base ten blocks, paper plates, coins and bills (or Master 8)

Introductory Activities

Using Models to Find the Pattern

Write on the board: Three children shared $4.62. How much did each get?

Use real or play coins and bills to show $4.62 with $1 bills, dimes and pennies. Ask one student to show how to share the money with three students and another student to record the sharing on the chalkboard.

Each student will receive $1.54.

Write on the board: A 4.62 kilogram box of raisins is to be packed into 3 smaller boxes, each holding the same amount. What is the weight of each box?

After students have decided that this problem is solved by division, build 4.62 with base ten blocks (4 flats, 6 longs and 2 units) and share the blocks among 3 paper plates, beginning with the largest size block. After the blocks have been shared, resulting in 1.54 blocks in each plate, reconstruct the manipulative activity as you demonstrate the traditional way to write the problem. Then, ask which operation is used to get the 1 in the quotient (divide), the 3 written under the 4 (multiply), the 1 left over (subtract) and the 16 (trading 1 flat for 10 longs) as you write D, M, S, ↓ (for trade or bring down) on the board.

$$\begin{array}{r} 1 \\ 3\overline{)4.62} \\ \underline{3} \\ 16 \end{array}$$

You have used the blocks to discover the four steps, which are repeated with each block: divide, multiply, subtract and bring down. You should continue to use blocks for division until you are certain you can repeat the four steps accurately.

You will want students to note that, when sharing models such as money or base ten blocks on paper plates, the blocks may be shared in any order; that is, you may share the biggest block first or the smallest block first, and the answer will still be the same. However, students will discover

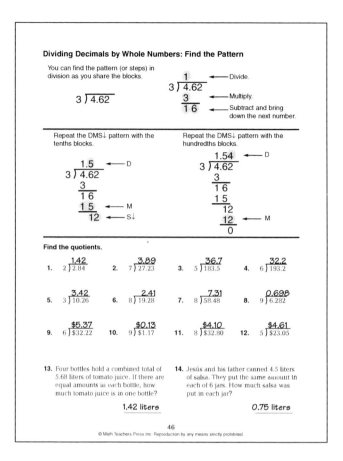

that the most efficient way is to share the biggest block first, and this method readily leads to the four steps in the division algorithm (divide, multiply, subtract, bring down). As students gain confidence in their own ability to invent arithmetic procedures, they eventually will decide that sharing the big block first is the most efficient.

About This Page

Use base ten blocks to demonstrate the problem at the top of the page. Build 4 flats for the ones, 6 rods for the tenths and 2 small cubes for the hundredths. Share each block on 3 paper plates as suggested.

Have students use base ten blocks to complete the first 3 problems and paper and pencil to complete the page.

Follow Up Activities

Skill Builders 28-1

Objective: To divide decimals by decimals.

Materials: Base ten blocks and paper plates

Introductory Activities

Dividing Decimals by Decimals

Write on the board: A board 1.8 meters long is to be cut into shelves 0.2 meters each. How many shelves can be cut from the board?

Ask students which problem-solving strategies they might use to solve this problem. (Draw a picture, use a model)

Have students draw a picture of the board to find the number of shelves.

Then build 1.8 with base ten blocks (1 whole, 8 tenths). Keep removing 0.2 (2 tenths blocks), until zero remains. (0.2 can be removed 9 times from 1.8) Show the recording:

$$0.2 \overline{)1.8} = 9$$

After the quotients have been written from the blocks, ask students to see if they can see a pattern for placing the decimal point, so they could work the problem without base ten blocks. Some of the students will begin verbalizing the patterns for moving the decimal point in the divisor.

At this point, explain that there are two different types of division problems, one where you know the number of groups and need to find how many are in each group (24 cookies shared among 3 students) and the other when you know how many are in each group but not the number of groups (boards .2 long to be cut from a board 1.8 long).

Explain that the easier way to divide is to use the first method with paper plates rather than removing sets of a given size. To use the paper-plates method, however, the divisor must be a whole number.

Write on the board:

$$\frac{1.8}{0.2} = \frac{1.8 \times 10}{0.2 \times 10} = \frac{18}{2} = 2\overline{)18}^{\,9}$$

How could I change .2 into the whole number 2? (multiply it by 10) **If we multiply the denominator by 10 what must we do to the numerator?** (multiply it by 10).

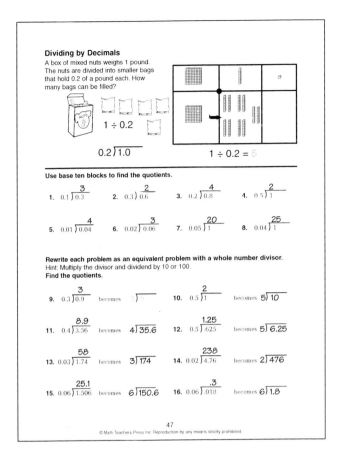

Write the problem again on the board, this time using arrows to show the movement of the decimal point that makes the divisor a whole number.

$$0.2 \overline{)1.8} = 9$$

About This Page

Read the illustration together. Solve using base ten blocks.

Problems 1–8 should be solved with base ten blocks as you ask, e.g., **How many 1 tenths are there in 3 tenths?**

Continue by having students complete problems 9–16.

Follow Up Activities

Skill Builders 28-2, 29-1, 30-1

Objective: To find a fractional part of a number.

Materials: Base ten blocks, string

Vocabulary: set, discount

Introductory Activities

Problem Solving: Acting Out the Problem

Today we are going to find the fractional part of a set. What do we mean by the word *set*? (a collection or a group of objects)

Have 18 students stand in front of the class. Write on the board: In a class of 18 students, one-half of them go to first lunch. How many go to first lunch?

How many students would go to the first lunch? (9) Ask a student to explain how they got the answer. (I divided 18 into 2 groups of equal size and there were nine in each group, or I thought about having 1 out of every 2 groups of students leave for lunch.)

Repeat with $\frac{1}{3}$ of 18 and $\frac{1}{6}$ of 18.

Problem Solving: Using a Model

Suppose a pizza is cut into 8 pieces. You came home from school yesterday and ate $\frac{1}{2}$ of the pizza. How many pieces did you eat? (4) **How did you get your answer?** Draw a picture of the pizza on the board. Have students discuss how to solve the problem. Students might draw a picture and use a pencil to show that $\frac{1}{2}$ of 8 pieces is 4 pieces.

After allowing the students to explain their answers, review the various methods for multiplying fractions by whole numbers.

Students could also use a series of ratios to find that 1 out of 2 slices is the same as 2 out of 4 slices is the same as 4 out of 8 slices of pizza.

Write on the board: $\frac{1}{2}$ of 8 = 4

What pattern would you use to solve this problem with paper and pencil? (Write 8 as a fraction $\frac{8}{1}$ and then multiply.)

$$\frac{1}{2} \times \frac{8}{1} = \frac{8}{2}$$

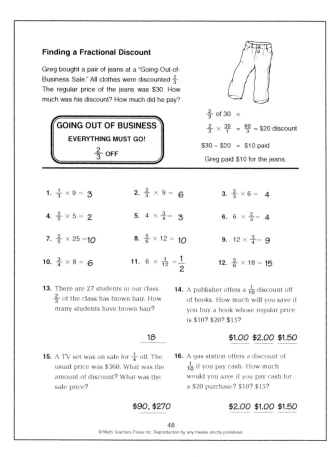

How many slices are in $\frac{3}{4}$ of a pizza? (6) Have students explain their solutions.

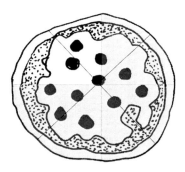

About This Page

Read the problem together at the top of the page. Ask students to give example of items they have bought at a discount. Count out 30 ones blocks. Ask a student to use a piece of string to show $\frac{1}{3}$ of them (10) and $\frac{2}{3}$ of them (20). Show the solution with paper and pencil. Work problems 1 and 4 with counters and string.

Have students complete the page on their own.

Follow Up Activities

Skill Builders 45-3

Objective: To find multiple prices and price per unit.

Materials: 2 boxes of cereal

Vocabulary: multiple price, price per unit

Introductory Activities

Using Food as Models

Display a box of cereal. Tell how much it weighs and how much it costs. Ask students to draw a picture to find how much 2 boxes would cost.

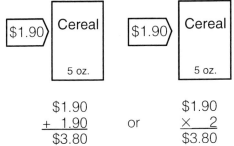

$1.90 $1.90
+ 1.90   or   ×   2
$3.80 $3.80

Write the answer in a complete sentence. (Two boxes of cereal cost $3.80.) **How much does 1 ounce of the cereal cost? Would it be more or less than $1.90** (less)

Draw a picture of the cereal box divided into 5 equal parts and label each part as 1 ounce.

```
        | 1 oz. |
        | 1 oz. |
$1.90   | 1 oz. |
        | 1 oz. |
        | 1 oz. |
```

Find the cost per ounce of the cereal by dividing the cost by the number of ounces in the box. Ask a student to explain and show the solution.

```
      0.38
   5)1.90
     15
     ‾‾
     40
     40
```

Display two different size boxes of the same cereal. Tell students how much each box weighs and costs. Ask which box is the better buy. If cereal boxes are not available, draw a picture of the following on the board:

Which is the better buy? How do you know? (The larger box is the better buy, because 1 ounce costs $0.38, but the small box costs $0.40 per ounce.)

About This Page

Work several problems together, asking a student to draw a picture of each problem to help decide on a process.

Follow Up Activities

Journal Prompt
Which is the better buy, a single candy bar for 45¢ or a package of 6 candy bars for $2.10? Use diagrams, words, and symbols to prove your answer is the better deal.

Skill Builders 45-4

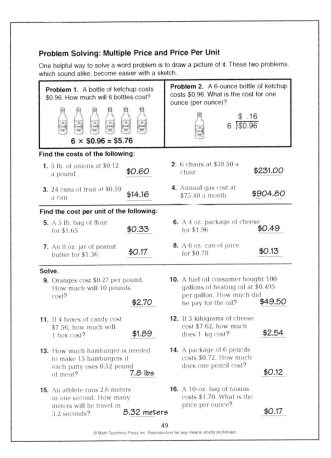

Objective: To identify lines as horizontal, vertical, diagonal, parallel, intersecting or perpendicular.

Materials: Geoboards or Geoboard Dot Paper (Master 12), masking tape, overhead geoboard (optional)

Vocabulary: horizontal, vertical, diagonal, parallel, intersecting, perpendicular

Introductory Activities

Kinds of Lines

Each small group or pair of students will need a geoboard and geobands.

Using a 25-peg geoboard, put a strip of masking tape below each row of pegs. Starting at the top left peg, write the letters A–E below each peg in the first row, F–J below each peg in the second row, K–O in the third row, P–T in the fourth row and U–Y in the bottom row as shown. Use an overhead pen to write the letters in similar positions on a 25-peg geoboard for the overhead. Use the overhead geoboard to parallel the activities with the students on their geoboards.

```
A  B  C  D  E
F  G  H  I  J
K  L  M  N  O
P  Q  R  S  T
U  V  W  X  Y
```

After demonstrating position words *horizontal*, *vertical* and *diagonal*, have students hold their arms in each position. Ask students to form a line on the geoboard that goes in each direction. **What are the names of the line segments that you formed?**

Write the different examples given by students on the board and ask the students to check to see if they agree. Have students demonstrate parallel lines and intersecting lines by holding their arms in different positions.

Form line segment FH on your geoboard. Form another line segment parallel to FH. Record student examples on the chalkboard and have the class check the responses.

Form line segment KO on your geoboard. Form another line segment intersecting line segment KO. Record student examples on the chalkboard and have the class check the responses.

Ask a student to draw a diagram of the classroom floor as viewed from above on the chalkboard. Use a piece of colored chalk to shade in each corner.

How are the two lines that form each corner related to each other? (The lines intersect and make a special square angle.) **We call lines that intersect to make square angles *perpendicular* lines. The two lines forming each corner are perpendicular.** Draw the box symbol in each of the colored corners of the figure drawn on the chalkboard.

Form line segment GJ on your geoboard. Form a line segment perpendicular to GJ. Record student examples and have the class check each response.

Ask students to identify other examples of square corners in the classroom (a sheet of paper, an index card, etc.)

About This Page

Read the example at the top of the page. Have students use a pencil to trace each different kind of line or lines. Have student volunteers give the answers to problems 1–4 and 15–16. Then have students complete the page on their own.

Follow Up Activities

Skill Builders 32-1

Objective: To measure and draw angles with a protractor.

Materials: Protractor (Master 9), scissors, chalkboard protractor (optional)

Vocabulary: protractor, degrees
Skill Builders Vocabulary: vertex

Introductory Activities

Measuring Angles

Find examples of right angles in the classroom (most corners, corner of a book or a piece of paper). Open a pair of scissors. Point to the shape made by the cutting edges and describe this shape as an example of an angle. Ask if the angle is more or less than the right angle found at the corners of a book. (less) **We call these angles acute angles.**

Ask students to put the palms of their hands together and slowly open their hands at the fingertips to form acute angles made by the cutting edge of a scissors. (The hands must remain together at the wrists.) **Show a right angle. Can you open your hands to show an angle greater than 90°?** Demonstrate. **We call angles greater than 90° but less than 180°** *obtuse* **angles.**

Draw right, acute and obtuse angles on the chalkboard. Before measuring each angle, ask students to estimate if the measure of each angle is less than, equal to or more than 90°. Place the protractor over the angle so that the center of the protractor is on the vertex and the 0° line on the straight edge lines up with one side of the angle. Move your finger along the arc of the protractor, counting up by 10 degrees to the other side of the angle.

Draw several angles in non-standard positions, i.e., where the initial side is not always drawn in horizontal position.

Drawing Angles

Pointing to the 0° mark on the right edge of the protractor, ask a student to estimate and use a straight edge to draw an angle of a given number of degrees on the chalkboard, e.g., an angle of 90°, 45°, 30°, 135°, 180°. Then have another student use the chalkboard protractor to draw the actual measurement of the angle next to the first student's estimate.

Discuss how close the estimated angles should be to the actual measurements to be considered

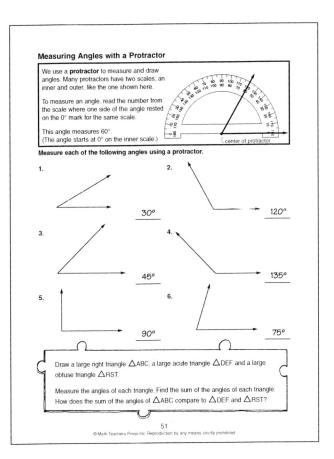

good estimates. (Estimates within 5° of the actual measurement would be very good guesses.)

Draw several angles on the chalkboard, both in standard and non-standard positions. Ask students to write an estimate of each angle on their paper. Ask a volunteer to measure each angle. An estimate within 5°–10° should be considered acceptable.

Ordering Angles

Draw and label 5 angles (A, B, C, D, E) on the chalkboard. Ask which angle is the smallest. Write a 1 under the smallest angle. Ask which is next smallest and write a 2 under that angle. Continue ordering the angles by writing 3, 4 or 5 under the remaining angles. Check estimates with a protractor.

About This Page

Read the example at the top of the page. Before measuring the angles in problems 1–6, have students estimate the measure of each angle.

Follow Up Activities

Skill Builders 31-1, 37-1

Objective: To identify angles as right, acute, obtuse or straight.

Materials: Geoboards (or Master 12), masking tape, overhead geoboards (optional)

Vocabulary: right, obtuse, acute, straight, congruent

Introductory Activities

Geoboard Activities

Using a 25-peg geoboard, put a strip of masking tape below each row of pegs. Starting at the top left peg, write the letters A–E below each peg in the first row, F–J below each peg in the second row, K–O in the third row, P–T in the fourth row and U–Y in the bottom row. Use an overhead geoboard to demonstrate and to parallel the activities with the students.

A	B	C	D	E
F	G	H	I	J
K	L	M	N	O
P	Q	R	S	T
U	V	W	X	Y

Have students find and label a pair of line segments that are the same length. **These line segments are congruent.** Show an angle on a geoboard. Name another angle that is congruent. Prove they are congruent by using Master 12 and cutting out the first angle and placing the cutout on the other angle.

Geoboard Angles

Draw a right angle HRT on the chalkboard or overhead geoboard. Have students form the angle on their geoboards. Describe ∠HRT. (A right angle, an angle with square corners, sides HR and RT are perpendicular.)

Draw an acute ∠JRT on the chalkboard. Have students form the same angle on their geoboards, using a contrasting color to ∠HRT.

How does ∠JRT compare to ∠HRT? (has a smaller measure) **Is ∠JRT more than, less than or equal to 90°?** (less than) **Estimate the measure of ∠JRT.** (45°) **Angles measuring less than 90° are called acute angles.**

Draw ∠GRT on the chalkboard and repeat the activity to identify obtuse angles as measuring more than 90°.

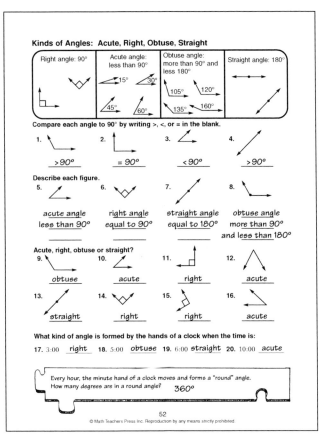

Draw ∠PRT on the chalkboard and repeat the activity to define a straight angle as two right angles of 90° each or 180°.

Using the circular side of the geoboard, have students form ∠BOD and describe the angle as acute, right, or obtuse. Repeat with ∠BOH (straight) and ∠BOF (obtuse).

Have students use toothpicks or straws to demonstrate angles equal to 90°, less than 90°, more than 90°, and equal to 180°.

About This Page

Ask students to study the two right angles drawn in the first illustration. **How does the size of the second angle compare to the size of the first?** (same) **How do you know?** (The small box always means 90°; the size of the angle does not change as the whole angle is rotated.)

Follow Up Activities

Journal Prompt

Think of the three types of angles: right, obtuse, and acute. Explain how these angles are different.

Skill Builders 33-1

Objective: To identify a polygon by the number of sides.

Materials: Toothpicks or straws, geoboards, Geoboard Dot Paper (Master 12), scissors, glue (optional), tagboard or construction paper (optional)

Vocabulary: open figure, closed figure, regular polygon, regular pentagon, regular hexagon, regular octagon, regular decagon, congruent, slides, flips, turns

Skill Builders Vocabulary: quadrilateral, triangle

Introductory Activities

Open and Closed Figures

Have students form capital letters of the alphabet on the geoboard to identify those that are open and closed. **Form the letter C on your geoboard. Trace the letter with your fingers, noticing where it begins and ends. Is the letter C an open or closed figure?** (open) Repeat with the letters D (closed), N (open) and Z (open).

Form the letter B on your geoboard. A simple closed figure comes back to where it started and never crosses itself. Is the letter B closed? (no) **Why?** (because it has a crossover)

A closed figure with four sides is called a quadrilateral. Is a rectangle a quadrilateral? (yes) Have students form a rectangle on the geoboard. **How would you describe a rectangle?** (A quadrilateral with 4 right angles; 2 long sides are parallel and equal; 2 short sides are parallel and equal.) Repeat with a square. (A quadrilateral with 4 right angles; opposite sides are parallel; all sides are equal.)

Naming Polygons

Use 5 toothpicks to form a polygon. **A polygon with 5 sides is a pentagon. How do the sides compare?** (equal) **A pentagon with equal sides is called a regular pentagon. Paste your polygon on your paper. Write "regular pentagon" under the figure.**

Repeat with 6 toothpicks (a regular hexagon), 8 toothpicks (a regular octagon) and 10 toothpicks (a regular decagon). Students may glue toothpick figures on tagboard, if desired.

About This Page

Read the illustration at the top of the page. Have students use a pencil to count and trace the

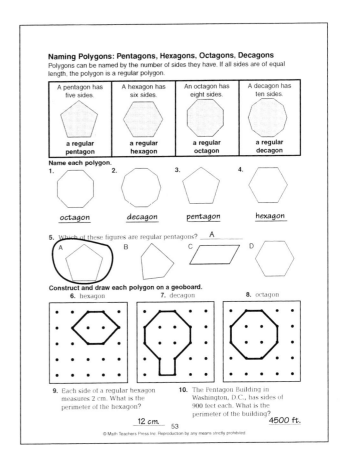

number of sides on each polygon. Problem 9 uses the word *perimeter*. Refer to page 56 to review the meaning of perimeter with the students.

Follow Up Activities

Slides, Flip, Turns

Give each student two copies of Master 12 and a pair of scissors. **Draw any 4-sided figure on the first sheet of dot paper. Cut out the figure.** Demonstrate how to place the cut-out figure on the second sheet of dot paper, slide it, and outline it in its new position. **How does the size of the polygon compare to the original?** (It is the same.) **We say it is congruent.**

Repeat the activity by flipping and rotating the cut-out figure on additional sheets of dot paper.

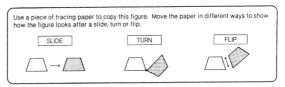

Skill Builders 34-1

6E Teacher Manual 53

Objective: To identify and draw the parts of a circle: center, radius, diameter, circumference.

Materials: Circular geoboard (Master 12), centimeter ruler (Master 9), chalkboard compass (or piece of chalk attached to a string), crayons, colored chalk, masking tape

Vocabulary: circle, center, radius, diameter, radii, circumference, compass, line segment

Introductory Activities

Naming Parts of a Circle

Have students use the circular side of the geoboard, writing letters A–L next to each point and O at the center on masking tape.

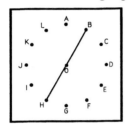

Outline a circle with one rubber band. **This rubber band shows the circumference of the circle. The center of the circle is at point O.** Put a small piece of paper on point O. Use a different colored rubber band and form a line from O to any point on the circumference. Form a line from O to a different point on the circumference and measure the lengths of each line. How do the lengths compare? (They are the same.) **Each line is called a radius. More than one radius are called radii. How many different radii can you form?** (12)

Form the radius OB on your geoboard. Extend the rubber band from point O in a straight line to the point directly opposite. (OB becomes HB.) **The line BH is a diameter of the circle. How does the diameter compare to the radius?** (twice as long) **How many diameters are in the circle?** (6) **Name them.** (AG, BH, CI, DJ, EK and FL)

Using a chalkboard compass (or a piece of chalk tied at the end of a string), demonstrate how to draw a circle of a given size, e.g., radius of 5 inches. Set the compass to measure 5 inches by first placing it on a ruler. Draw and shade the radius and diameter, using different colored chalk.

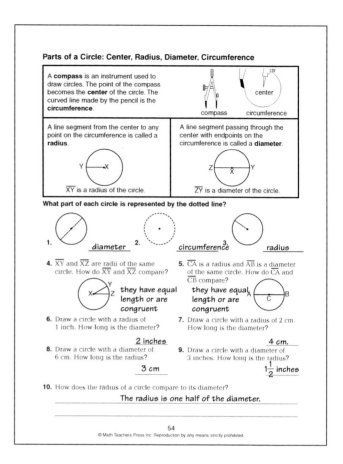

About This Page

Read the example at the top of the page. Have students use a pencil to trace the circumference, radius and diameter in the example.

Answer problems 1, 4, and 6 together.

Have students complete the page on their own.

Follow Up Activities

Skill Builders 35-1

Objective: To measure and draw lines to the nearest $\frac{1}{4}$ inch.

Materials: Tagboard strips, *Skill Builders* 36-1, scissors, glue, ruler or Master 9

Introductory Activities

Making a Ruler to the Nearest Half-Inch

Each student will need scissors, glue, and a copy of *Skill Builders* 36-1. Have students make a ruler divided into $\frac{1}{2}$ inches. Then fold a $\frac{1}{2}$-inch piece of tagboard into two smaller matching parts and identify each part as $\frac{1}{4}$ inch. Have students mark $\frac{1}{4}$-inch intervals on the ruler and write a fraction next to each mark.

About This Page

Have students count by $\frac{1}{2}$s from $\frac{1}{2}$ to $5\frac{1}{2}$, touching each point on the ruler, marking over the line as it is read. Then have students count by $\frac{1}{4}$s from $\frac{1}{4}$ to $5\frac{1}{2}$ inches. ($\frac{1}{4}$, $\frac{2}{4}$, $\frac{3}{4}$, $\frac{4}{4}$ or 1, $\frac{5}{4}$ or $1\frac{1}{4}$, $\frac{6}{4}$ or $1\frac{2}{4}$, $\frac{7}{4}$ or $1\frac{3}{4}$, … $\frac{21}{4}$ or $5\frac{1}{4}$, $\frac{22}{4}$ or $5\frac{2}{4}$ or $5\frac{1}{2}$) **Write $\frac{1}{4}$ at the correct mark on your ruler on the page. Now put your pencil on the mark that is $\frac{1}{2}$ of $\frac{1}{4}$.** The smallest marks show $\frac{1}{2}$ of $\frac{1}{4}$ inch. **How much is $\frac{1}{2}$ of $\frac{1}{4}$?** ($\frac{1}{8}$) **Count by $\frac{1}{8}$s with me as you touch each mark along the ruler from 0 to 1.** ($\frac{1}{8}$, $\frac{2}{8}$ or $\frac{1}{4}$, $\frac{3}{8}$, $\frac{4}{8}$ or $\frac{1}{2}$, $\frac{5}{8}$, $\frac{6}{8}$ or $\frac{3}{4}$, $\frac{7}{8}$, $\frac{8}{8}$ or 1)

Have students use a ruler cut from Master 9 to complete the page.

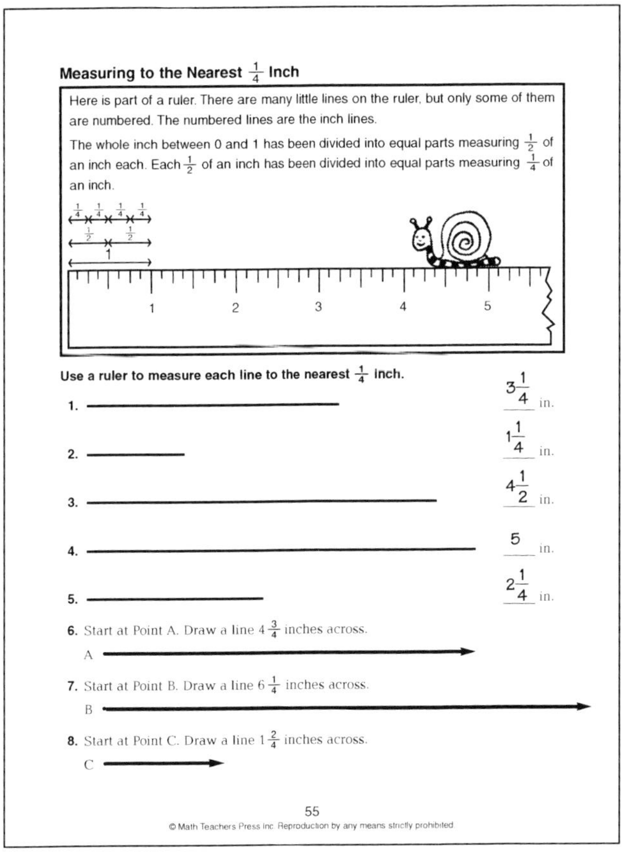

Measuring to the Nearest $\frac{1}{4}$ Inch

Here is part of a ruler. There are many little lines on the ruler, but only some of them are numbered. The numbered lines are the inch lines.

The whole inch between 0 and 1 has been divided into equal parts measuring $\frac{1}{2}$ of an inch each. Each $\frac{1}{2}$ of an inch has been divided into equal parts measuring $\frac{1}{4}$ of an inch.

Use a ruler to measure each line to the nearest $\frac{1}{4}$ inch.

1. $3\frac{1}{4}$ in.
2. $1\frac{1}{4}$ in.
3. $4\frac{1}{2}$ in.
4. 5 in.
5. $2\frac{1}{4}$ in.
6. Start at Point A. Draw a line $4\frac{3}{4}$ inches across.
7. Start at Point B. Draw a line $6\frac{1}{4}$ inches across.
8. Start at Point C. Draw a line $1\frac{2}{4}$ inches across.

55

Follow Up Activities

Drawing Lines

Have students use a ruler cut from Master 9 to draw lines of different lengths: 2 inches, $3\frac{1}{2}$ inches, $1\frac{1}{4}$ inches, $4\frac{1}{8}$ inches.

Skill Builders 36-2

Objective: To find the perimeter of polygons using objects. **DVD**

Materials: Units blocks, Centimeter Ruler (or Master 9), geoboards (or Master 12)

Vocabulary: perimeter

Introductory Activities

Perimeter of Polygons

Display the outline of a 3 cm × 4 cm rectangle. Draw flowers inside the rectangle.

Here is a picture of Sally's garden. She wants to place bricks around the outer edge to separate the garden from the rest of the yard. Each brick is the size of this unit block. How many bricks will she need?

Display a unit block. Ask students to estimate how many bricks will be needed to go around the outside of the garden. Then ask a student to physically place the blocks around the outside to find the actual number needed. (14 blocks)

What part of the rectangle have we measured? (the distance around the outside) **We call the distance around the outside of a figure its perimeter. What is the perimeter of the garden in bricks?** (14)

Repeat finding the perimeters of other polygons having sides of a whole number of centimeters. Watch to see that students do not place a cube in each of the 4 corners as the corners are not part of each side.

Display the outline of a 2 cm × 6 cm rectangle on the overhead. Review the definition of perimeter. Estimate the perimeter to the nearest centimeter. Record different estimates on the chalkboard to refer to later.

How can we find the actual perimeter of this rectangle if we only have one block to use? (Mark off units of 1 block each along the sides and add the number of spaces together.)
How many units on the longest side? (6)
How many units on the shortest side? (2)
How many units along the perimeter? (16)
Have a student volunteer mark off units along the sides of the rectangle and explain his or her answer. Compare the actual perimeter to the estimated perimeters.

Repeat with other examples.

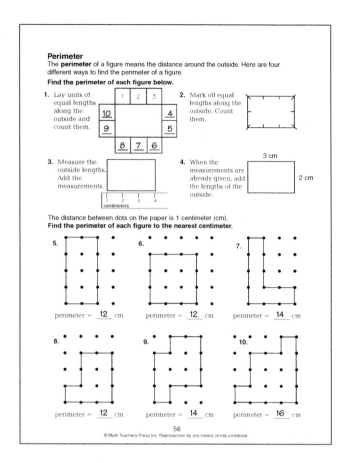

Estimate the number of feet in the perimeter of the classroom. Have a student find the perimeter of the classroom by walking and counting paces. Have another student find the perimeter with rulers placed end to end. Compare the actual perimeter to the estimated perimeters.

About This Page

Demonstrate each of the four different ways to find a perimeter given at the top of the page. Have students complete the page on their own.

Follow Up Activities

Journal Prompt

Sam measured the outside of his vegetable garden. The sides measured 6 feet, 3 feet, 6 feet and 3 feet. How much fencing does Sam need to put a fence around the whole garden? Show two different methods for finding the solution.

Skill Builders 38-1

Objective: To find the area of polygons using square units. **DVD**

Materials: Centimeter squares cut from Centimeter Graph Paper (Master 2)

Vocabulary: area, square

Introductory Activities

Exploring Shapes for Measuring Area

Each student will need a 3 cm × 5 cm rectangle, a sheet of Centimeter Graph Paper (Master 2) and scissors. Display an outline of a 3 cm × 5 cm rectangle.

Cut out several rows of 1-centimeter squares from a sheet of centimeter graph paper. **About how many of these small squares will cover the rectangle?** Record the estimates. **Cover the surface of the rectangle with the 1-centimeter squares. How many squares will completely cover the surface?** (15) **We say the area is 15** *square* **centimeters.**

We have been finding the area of the rectangle. Can you define area in your own words? (Finding the number of square units needed to cover a surface.)

Outline and cut out a rectangle that has an area of 12 square centimeters. Display different rectangles drawn by class members to confirm that each rectangle has an area of 12 square centimeters. (There are 3 possible: 1 × 12, 2 × 6 and 3 × 4.)

What do you notice about these rectangles which have an area of 12 square centimeters? (They have different shapes.)

About This Page

Read the example at the top of the page. Have students fold centimeter squares on the diagonal to cover the small triangular parts in each corner in problem 4.

Follow Up Activities

Same Areas, Different Perimeters

Have students find the perimeter of different rectangles having a common area of 12 square centimeters. (Perimeters of a 1 × 12 = 26;

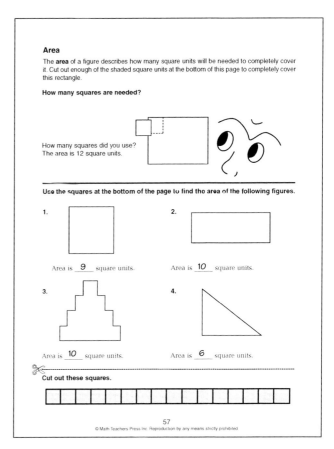

2 × 6 = 16; 3 × 4 = 14 cm) **What do you notice about the perimeters of different figures that have the same area?** (They have different perimeters.)

The Square of a Number

Draw a picture of a square with 3 units on each side.

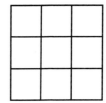

This is a picture of the square of three. There are 9 small squares covering the square of three. We can write this in two ways. Write on the board; $3 \times 3 = 9$ or $3^2 = 9$.

Find the value of:
1. the square of 4 (16)
2. 5 squared (25)
3. 6^2 (36)
4. 10^2 (100)
5. 8^2 (64)

Skill Builders 38-2

Objective: To find the volume of a one-layer rectangular solid. **DVD**

Materials: Units blocks, Centimeter Graph Paper (Master 2), scissors, tape

Vocabulary: volume, capacity

Skill Builders Vocabulary: edge, face

Introductory Activities

Measuring Volume

Measuring volume can be approached either by counting the cubic units inside a solid figure or pouring a substance into a measuring container to see what portion of the container it fills. This page focuses on the first approach.

Making One-Layer Boxes

Use centimeter graph paper to make a box (without a cover) **that is 3 cm long, 2 cm wide and 1 cm high. Fill the box with cubic unit blocks to find its volume.** (6 cubes or 6 cu. cm)

Repeat with a 4 × 3 and a 3 × 5, each 1 cm high (volumes of 12 cu. cm and 15 cu. cm). **Volume is the number of cubes needed to fill a box.**

About This Page

Read the illustration together and relate the 1 cm × 1 cm × 1 cm cube with a unit block. One unit block has a volume of 1 cubic centimeter. **What is the volume of 2 unit blocks?** (2 cu. cm)

Many students have difficulty translating from a 2-dimensional picture to a 3-dimensional solid. When asked to find the volume of the picture in problem 1, many students say "7" as they count the faces they see rather than the actual cubes.

Have students work with a partner. One person uses the unit blocks to match the rectangular solid, the other person counts the cubes to find the volume and writes the answer with "cu. cm."

Follow Up Activities

Surface Area

Have the students cut out centimeter squares to find the surface area of each of the cubes. (Students cover each face with squares.)

The Cube of a Number

Use cubes to demonstrate the cube of a number. The cube of 2 means to build a cube with 2 units on each edge. **How many small cubes are in the cube of 2?** (8)

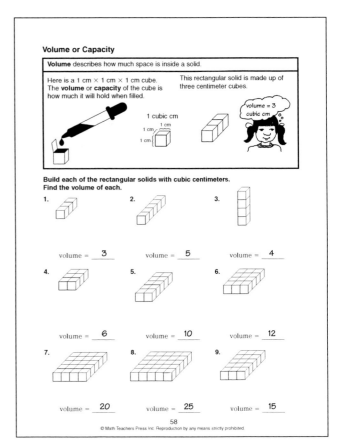

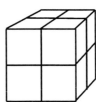

Write on the board:
   The cube of 2 or "two cubed" is 8.
   $2 \times 2 \times 2 = 8$ or $2^3 = 8$

Write on the board: Use cubes to find
1. the cube of 3 (27)
2. 5^3 (125)
3. 4 cubed (64)
4. 10^3 (1000)
5. 6^3 (216)

 Journal Prompt
   Which type of measurement (perimeter, area, volume) would you use to find
1. the amount of carpet for a bedroom?
2. the amount of water in an aquarium?
3. the amount of blocks that fit in a box?
4. the amount of fencing for a garden?

Skill Builders 39-1, 39-2, 42-1

Objective: To find the interval between two given times.

Materials: Two clocks with moveable hands

Introductory Activities

Review Time to the Nearest 5 Minutes

Display an overhead clock showing 2 o'clock. Move both hands around to show 3 o'clock. **Over how much of the clock did the minute hand turn?** (All the way around the clock.) **This shows that one whole hour or 60 minutes has passed.**

Move the hands to show 3:30. **Over how much of the clock did the minute hand turn?** (It turned halfway around the clock.) **This shows that half an hour has passed. We say the time is half past the hour of 3 or 30 minutes after 3 or 3:30.**

Repeat with other "half-past" times. Ask students to describe the pattern for the hour hand and the minute hand for the "half-past" time. (The hour hand points between two numbers and the minute hand always points to 6.)

Display an overhead clock showing 4:15. **Where is the hour hand pointing?** (a little after the 4) **Is the time before or after 4 o'clock?** (after) **About how much after 4 o'clock do you think it is?** (Answers will vary.)

How many minutes between 4 o'clock and 5 o'clock? (60) **How many minutes to go around the whole clock?** (60)

Where is the minute hand pointing? (to the 3) **How many minutes after 4 is it?** (15) **How did you get your answer?** (3 skip counts of 5) **We say it is 15 minutes after 4 or 4:15.**

Repeat with a clock set at 4:30, 4:45, 5:10, 5:25, etc.

Display 8:25 on the clock. **What time is it?** (8:25) **Move the minute hand two minutes past 8:25. What time is it?** (8:27) **How did you get your answer?** (Added 2 minutes to 25 minutes.)

Repeat with 8:39, 8:51, etc.

About This Page

Display two clocks with moveable hands, one set at the beginning time of 8:30 a.m. and the other set at the ending time of 3:00 p.m. **To find

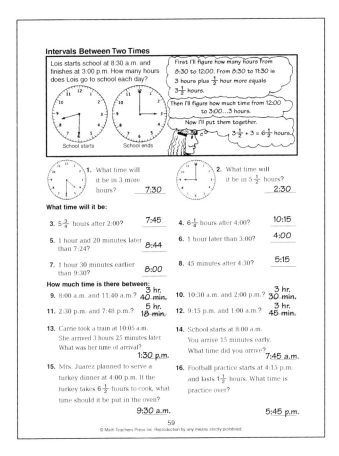

the amount of time at school, think of moving the hands on the first clock to match the hands on the second clock.**

How many minutes as the minute hand moves from 6 to 12 and the hour hand moves to 9? (30 minutes) **How many hours from 9:00 12:00?** (3 hours) **How many hours from 12:00 to 3:00?** (3 hours) **How many hours in all?** (6 hours 30 minutes)

Follow Up Activities

Skill Builders 40-1

Objective: To estimate and convert customary units of weight.

Materials: A full 1-pound coffee can, a stick of gum, base ten blocks, a primer balance (optional)

Vocabulary: ounces or oz., pounds or lbs., ton, weight

Introductory Activities

Estimate the Heavier or Lighter Object

Hold up pairs of classroom objects, e.g., an eraser and a book. **Which weighs more?** Have children take turns lifting each object to help them estimate. If a primer balance is available, refer to the balance with empty buckets. **What is the position of both buckets?** (They are even or balanced.) **What will happen to the buckets if I put the eraser in one and the book in the other?** (The bucket with the book stays down because the book weighs more.)

Display a full 1-pound can of coffee. **This can of coffee weighs 1 pound.** Point to the label. **There are 16 ounces in 1 pound.** Write on the board: 16 ounces (oz.) = 1 pound (lb.)

What do you buy in ounces or pounds? (candy, nuts, butter, flour, etc.) **A stick of gum weighs about 1 ounce.** Display 1 stick of gum.

About This Page

Follow these 4 suggestions to help students convert measurements:
1. Display actual examples of each unit being converted whenever possible.
2. Draw a picture to help decide how to solve a problem. Draw a picture of 1 pound of butter and show 16 ounces inside the outline.
3. Relate the pattern for whole number multiplication and division as you talk about converting with a table.
4. Complete a table and look for the pattern.

        1 lb.   16 oz.
        2 lb.   32 oz.
        3 lb.   48 oz.

The pattern: multiply the number of pounds by 16.

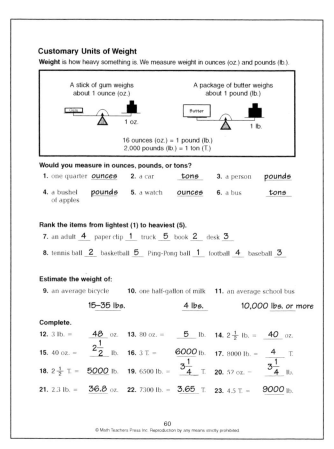

Follow Up Activities

Skill Builders 41-1

Objective: To graph ordered pairs on a coordinate grid (first quadrant only).

Materials: Masking tape or chalk, tagboard

Vocabulary: coordinate points, ordered pair

Introductory Activities

Graphing Exercise

Before the students arrive for class, put coordinate axes on the floor of the classroom using masking tape or chalk. (You may also use an overhead transparency and just project the setting on the screen.) Mark a corner (0, 0) as the starting point. Be sure the desks are arranged in columns and rows so that the coordinates of the position of each desk is evident.

1. Pass out cards with an ordered pair indicated to each student as they arrive. The seating arrangement is determined by the coordinates they receive. Be sure that the point on each card is unique.
2. Ask the students to stand, if the coordinates of their desk satisfy the following conditions:
    a. The point is on the x axis
    b. The point is in the first quadrant
    c. The desk is at point (2, 3)
    d. Satisfies the equation: $y = x$
    e. Satisfies the inequality: $y < x$
    f. Satisfies the equation: $y = x^2$

Students may memorize the instruction "over and up" to remember the correct order for naming coordinate points.

About This Page

Read the illustration and answer the questions together. Point B is at (3, 1); Point C is at (4, 5); Point D is at (6, 3); Point E is at (7, 2).

How many blocks would you have to walk from Point B to Point C? Is there more than one way to get from Point B to Point C? How many different ways can you find? Repeat the same questions about the distance between Point B and Point D. Emphasize that they must walk on the streets or avenues, not through or across the blocks.

After students see that a straight line is found when the coordinate points are connected, in problems 1 and 2, ask students if they think there

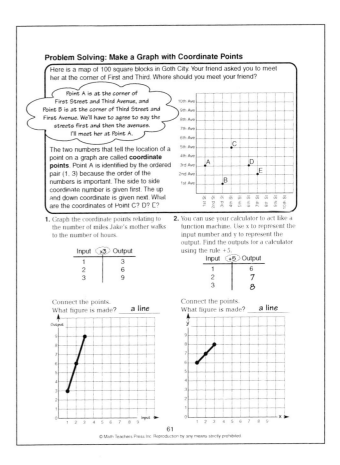

are any other coordinate points that would also be on this line. (4, 12) Have students suggest another point and show that each pair will be on the straight line.

Follow Up Activities

Graphing in the Second Quadrant

Draw a picture of a thermometer from −10° to +10° on a piece of tagboard. Relate each mark to temperature above and below zero. **The temperature is 5°. The temperature goes down 15°, what is the temperature?** Place a 10 × 10 grid along both parts of this number line. Demonstrate graphing (−2, 1) by starting at 0, counting 2 to the left and 1 up.

Objective: To collect data about favorite foods and represent the data in different ways.

Materials: Index cards, tape, Inch Graph Paper (Master 7)

Vocabulary: pictograph, bar graph

Introductory Activities

Collecting Data

Each student should have an index card and crayons. Students are interested in real data about themselves and their world. In this activity students use cards to collect data about their favorite pizza and describe the information using different pictorial representations (picture cards) and abstract representations (tables and graphs). It is important to have students choose how to investigate the data and discuss how to organize it.

Ask students to tell what the word favorite means. (Something that is liked best.) **What is your favorite pizza? Draw a picture of it on an index card.**

How can we organize this information so we can see the favorite pizza of all the children in the class? (Students may suggest having all the students with the same favorite food get into a group.) Have them move around the room to find children with matching pictures.

Look around the room at the different groups. What is the favorite pizza for the class? How do you know? Ask other questions about the groups so students discover the difficulty of using data organized this way.

We are finding interesting things about favorite pizzas, but sometimes it is hard to see everything when the people are standing in groups. How could we organize this information in a better way, so that we could all see it? (sort the food pictures) Allow students to sort food pictures into matching piles. Tape matching pictures in columns on the chalkboard or a large sheet of paper. **What does this data tell us?**

The data taped on the wall shows a picture of how many students chose each food as their favorite. This is an example of a picture graph or pictograph. There are other ways to show this information. We could use tally marks. Show how to make tallies in groups of 5.

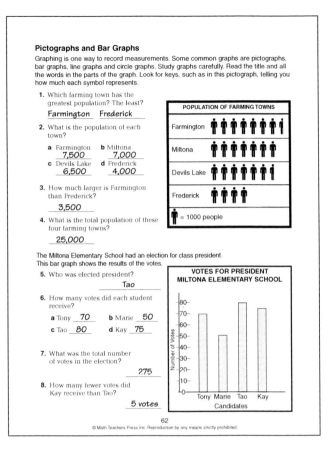

We could also show this as a table.

Favorite Pizzas	Number of Students
cheese	6
pepperoni	7
sausage	3
veggie	4

We can also make a bar graph to see everything at once. You can make a bar graph from the graph paper. First draw a picture or write the name of each favorite pizza at the bottom. Then shade 1 bar or space upward for each picture or tally mark.

About This Page

Help students learn to be good readers of graphs and tables by asking questions about details in the graph. **What is the title of the graph? Why is this graph called a pictograph? What does each stick figure represent?** (1000 people) **How much would 2 stick figures represent?** (2000 people)

Problem 3 on the pictograph and problem 4 on the bar graph will be the most difficult for the students. Remind them that to find how much larger one town is than the other, they can draw lines to match the stick figures from each town and then count the unmatched figures.

62   6E Teacher Manual

Objective: To read and interpret line graphs.

Materials: Graph paper (Master 2 or Master 7)

Vocabulary: line graph

Introductory Activities

Making Line Graphs

Scores on daily reviews or on timed tests of basic facts may be used to make a line graph showing progress for each student. Each student will need a copy of Master 2 or Master 7 to show the data. Some general rules to remember in making graphs include:
1. Every graph must have a title.
2. The parts of a graph must clearly show
   a. a name for each part or axis
   b. how many units that name represents (i.e., weeks, hours, etc)

If test scores are not available, write the following on the chalkboard:

Mark's daily math scores are:
- Test 1: 5 correct
- Test 2: 3 correct
- Test 3: 6 correct
- Test 4: 4 correct
- Test 5: 7 correct
- Test 6: 9 correct
- Test 7: 10 correct

Make a line graph showing Mark's daily math scores. Give your graph a title and label the horizontal and vertical axis.

About This Page

Before students answer questions related to each graph, ask, **What is the title of the graph? What information is given on the horizontal axis of the graph? What information is given on the vertical axis?**

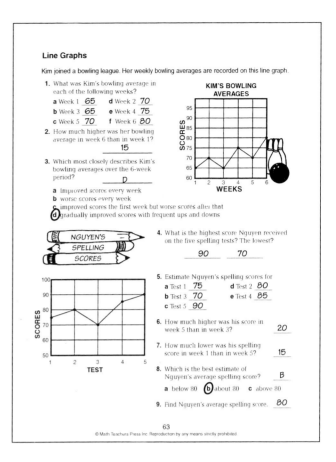

Follow Up Activities

Journal Prompt

Which kind of graph would you use to display:
- the week's spelling test scores for the class?
- six weeks of spelling test scores for one person?

Explain why you would use the graph you choose.

Skill Builders 48-1

6E Teacher Manual   63

Objective: To solve 1- and 2-step word problems involving reading tables.

Materials: Local fast-food menu, order pads, calculator, Place Value Mats (Masters 10 and 11), coins and bills (Master 8)

Skill Builders Vocabulary: probability

Introductory Activities

Reading Tables

Have students work in pairs with a copy of a menu from a local fast food restaurant and several sheets of order pads. Have 1 person order a lunch of 2–4 items and the other person record the order, the amount of each item and the total bill.

Two-step word problems are difficult because there is an intermediate problem that must be solved. To assure understanding of the two-step process, have students act out the problem with a waiter or waitress and a customer. **Example: Here are 2 sodas. Give me the money you owe me. Here are 3 fries. Give me the money you owe me. How much money do you owe me in all?**

Multiplying Money

Write on the board: Find the cost of 3 apples at 25¢ each. Use real or play money to build 3 groups of 25¢ with dimes and pennies on a Place Value Mat. Put together the pennies and dimes; exchange 10 pennies for 1 dime. Ask a student to write the problem on the board and show the regrouping.

Then show the solution of 3×0.25 with base ten blocks. Have a student write the problem in vertical format on the chalkboard and explain the regrouping of 10 hundredths to 1 tenth in the recording of the product.

About This Page

Read the menu at the top of the page. Discuss the 3 steps in solving the problem as you write each step on the chalkboard.

Step 1: $2 \times 0.80 = 1.60$

Step 2: $2 \times 0.40 = 0.80$

Step 3:
$$\begin{array}{r} 1.60 \\ +0.80 \\ \hline \$2.40 \end{array}$$

Have students follow a similar pattern for each problem.

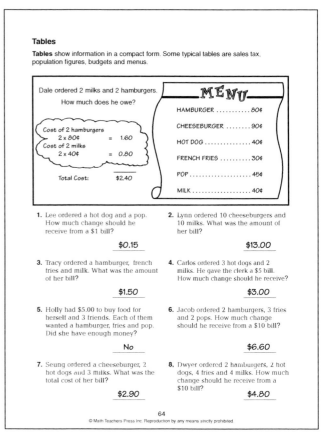

Follow Up Activities

Skill Builders 47-1, 47-2

Skill Builders

The *Skill Builders* section contains masters and reteaching pages. Copies of these masters should be made in advance.

Masters

Master 1	Place Value Bingo
Master 2	Centimeter Graph Paper
Master 3	100 Basic Addition Facts
Master 4	100 Basic Subtraction Facts
Master 5	100 Multiplication Facts
Master 6	90 Division Facts
Master 7	Inch Graph Paper
Master 8	Coins and Bills
Master 9	Rulers and Protractor
Master 10	Place Value Mat (Part 1)
Master 11	Place Value Mat (Part 2)
Master 12	Geoboard Dot Paper
Master 13	Glossary
Master 14	Mathematics Table
Master 15	Vocabulary Cards

The *Skill Builders* pages provide flexibility to accommodate any length of extended school. *Skill Builders* reteaching pages are numbered at the bottom from 1-1 to 50-2. For example, *Skill Builders* 19-2 teaches objective 19 and is the second page for reteaching that objective.

The inside back cover of the student book has a list of *Skill Builders* and a place to record when students complete each page. Students may ask you for *Skill Builders* pages by noting which objectives they missed on the daily reviews.

Place Value Bingo

Card 1

Hundreds	Tens	Ones

Card 2

Hundred Thousands	Tens Thousands	Ones Thousands

Card 3

Hundred Millions	Tens Millions	Ones Millions

Card 4

Tenths	Hundredths	Thousandths

1. Students write each of the digits 1–9 in random order on the squares on each card.
2. Teacher prepares 27 index cards for each of the 4 cards as follows:
   Card 1: 1 one, 2 ones, … 9 ones; 1 ten, 2 tens, … 9 tens; 1 hundred, 2 hundreds, … 9 hundreds.
   Card 2: 1 one thousand, … 9 one thousands; 1 ten thousand, … 9 ten thousands; 1 hundred thousand, … 9 hundred thousands.
   Card 3: 1 one million, … 9 one millions; 1 ten millions, … 9 ten millions; 1 hundred millions, … 9 hundred millions.
   Card 4: 1 tenth, … 9 tenths; 1 hundredth, … 9 hundredths; 1 thousandth, … 9 thousandths.
3. Teacher (or student) selects and reads aloud 1 index card at a time, e.g., "6 hundreds" from Card 1. Every player having a 6 written in one of the hundreds squares covers the space. The first player to have 3 in a row in any direction is the winner.

Master 1
© Math Teachers Press, Inc.
Reproduction only for one teacher for one class.

Centimeter Graph Paper

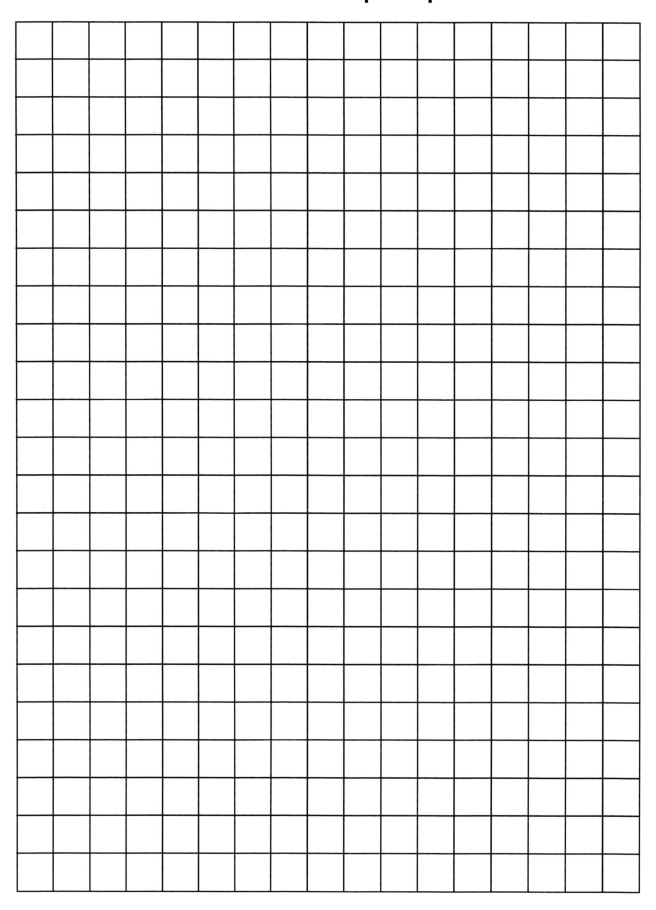

Master 2
© Math Teachers Press, Inc.
Reproduction only for one teacher for one class.

Addition Facts

100 Basic Addition Facts   Score _____

5 + 3	8 + 1	2 + 5	9 + 6	7 + 9	1 + 2	6 + 7	3 + 0	4 + 4	0 + 8
1 + 7	7 + 6	4 + 3	5 + 8	8 + 2	6 + 4	0 + 5	9 + 9	2 + 0	3 + 1
0 + 2	2 + 9	7 + 7	9 + 3	4 + 0	3 + 5	6 + 8	5 + 6	1 + 1	8 + 4
4 + 8	8 + 7	6 + 9	7 + 0	9 + 5	5 + 1	0 + 4	1 + 3	3 + 6	2 + 2
7 + 4	3 + 3	2 + 1	8 + 9	6 + 0	9 + 8	1 + 6	4 + 5	0 + 7	5 + 2
6 + 1	2 + 4	0 + 6	3 + 2	7 + 3	9 + 7	5 + 0	8 + 8	4 + 9	1 + 5
5 + 4	7 + 2	6 + 5	1 + 9	0 + 1	8 + 3	2 + 8	3 + 7	9 + 0	4 + 6
8 + 6	3 + 9	9 + 4	6 + 2	5 + 5	1 + 0	0 + 3	1 + 1	7 + 8	2 + 7
0 + 0	8 + 5	1 + 8	6 + 6	2 + 3	4 + 7	7 + 1	9 + 2	3 + 4	5 + 9
7 + 5	3 + 8	4 + 2	8 + 0	2 + 6	0 + 9	9 + 1	6 + 3	5 + 7	1 + 4

Master 3

Subtraction Facts

100 Basic Subtraction Facts Score _____

7 − 1	13 − 6	9 − 4	16 − 7	6 − 6	12 − 4	7 − 7	4 − 1	5 − 1	9 − 8
11 − 7	6 − 3	8 − 4	4 − 3	10 − 6	16 − 9	7 − 6	6 − 5	14 − 8	11 − 2
6 − 3	14 − 9	9 − 9	7 − 2	17 − 8	2 − 1	4 − 2	14 − 5	11 − 5	10 − 8
6 − 2	15 − 7	3 − 2	8 − 5	15 − 8	7 − 5	8 − 1	13 − 8	2 − 2	5 − 2
12 − 8	10 − 9	10 − 4	10 − 1	12 − 5	3 − 2	10 − 5	16 − 8	5 − 5	10 − 2
9 − 6	1 − 0	0 − 0	9 − 1	15 − 9	11 − 8	13 − 5	11 − 6	9 − 5	12 − 9
5 − 3	8 − 0	13 − 7	4 − 4	7 − 0	2 − 0	8 − 7	3 − 0	10 − 7	5 − 0
8 − 6	12 − 7	10 − 3	14 − 6	12 − 6	9 − 0	9 − 7	5 − 4	9 − 3	7 − 4
4 − 0	17 − 9	7 − 3	112 − 3	13 − 9	3 − 3	15 − 6	6 − 4	6 − 1	14 − 7
11 − 4	8 − 3	18 − 9	11 − 3	11 − 9	13 − 4	9 − 2	3 − 1	3 − 3	1 − 1

Multiplication Facts

100 Basic Multiplication Facts   Score _____

5	8	2	9	7	1	6	3	4	0
×3	×1	×5	×6	×9	×2	×7	×0	×4	×8

1	7	4	5	8	6	0	9	2	3
×7	×6	×3	×8	×2	×4	×5	×9	×0	×1

0	2	7	9	4	3	6	5	1	8
×2	×9	×7	×3	×0	×5	×8	×6	×1	×4

4	8	6	7	9	5	0	1	3	2
×8	×7	×9	×0	×5	×1	×4	×3	×6	×2

7	3	2	8	6	9	1	4	0	5
×4	×3	×1	×9	×0	×8	×6	×5	×7	×2

6	2	0	3	7	9	5	8	4	1
×1	×4	×6	×2	×3	×7	×0	×8	×9	×5

5	7	6	1	0	8	2	3	9	4
×4	×2	×5	×9	×1	×3	×8	×7	×0	×6

8	3	9	6	5	1	0	4	7	2
×6	×9	×4	×2	×5	×0	×3	×1	×8	×7

0	8	1	6	2	4	7	9	3	5
×0	×5	×8	×6	×3	×7	×1	×2	×4	×9

7	3	4	8	2	0	9	6	5	1
×5	×8	×2	×0	×6	×9	×1	×3	×7	×4

Division Facts

100 Basic Division Facts Score _____

$4\overline{)8}$ $2\overline{)0}$ $7\overline{)49}$ $5\overline{)30}$ $6\overline{)12}$ $9\overline{)9}$ $3\overline{)24}$ $8\overline{)56}$ $1\overline{)6}$

$8\overline{)40}$ $4\overline{)36}$ $6\overline{)30}$ $3\overline{)12}$ $5\overline{)25}$ $1\overline{)8}$ $7\overline{)0}$ $9\overline{)27}$ $2\overline{)8}$

$6\overline{)48}$ $9\overline{)18}$ $3\overline{)0}$ $5\overline{)45}$ $1\overline{)7}$ $2\overline{)4}$ $8\overline{)16}$ $7\overline{)63}$ $4\overline{)4}$

$7\overline{)56}$ $1\overline{)4}$ $4\overline{)16}$ $9\overline{)72}$ $8\overline{)64}$ $6\overline{)6}$ $5\overline{)15}$ $2\overline{)12}$ $3\overline{)18}$

$1\overline{)3}$ $8\overline{)48}$ $4\overline{)24}$ $7\overline{)7}$ $6\overline{)42}$ $9\overline{)54}$ $3\overline{)9}$ $2\overline{)16}$ $5\overline{)0}$

$9\overline{)36}$ $5\overline{)40}$ $1\overline{)2}$ $8\overline{)0}$ $3\overline{)3}$ $4\overline{)20}$ $2\overline{)10}$ $7\overline{)28}$ $6\overline{)54}$

$5\overline{)35}$ $2\overline{)2}$ $9\overline{)45}$ $6\overline{)36}$ $7\overline{)21}$ $1\overline{)0}$ $3\overline{)21}$ $4\overline{)12}$ $8\overline{)24}$

$2\overline{)14}$ $3\overline{)15}$ $7\overline{)35}$ $1\overline{)9}$ $5\overline{)10}$ $4\overline{)32}$ $6\overline{)0}$ $8\overline{)32}$ $9\overline{)63}$

$3\overline{)6}$ $4\overline{)28}$ $8\overline{)8}$ $2\overline{)18}$ $9\overline{)0}$ $4\overline{)20}$ $1\overline{)1}$ $6\overline{)24}$ $7\overline{)42}$

$4\overline{)0}$ $7\overline{)14}$ $6\overline{)18}$ $5\overline{)5}$ $8\overline{)72}$ $2\overline{)6}$ $3\overline{)27}$ $9\overline{)81}$ $1\overline{)5}$

Master 6

Inch Graph Paper

Master 7
© Math Teachers Press, Inc.
Reproduction only for one teacher for one class.

Coins and Bills

Master 8

Rulers and Protractor

Master 9
© Math Teachers Press, Inc.
Reproduction only for one teacher for one class.

Place Value Mat (Part 1)

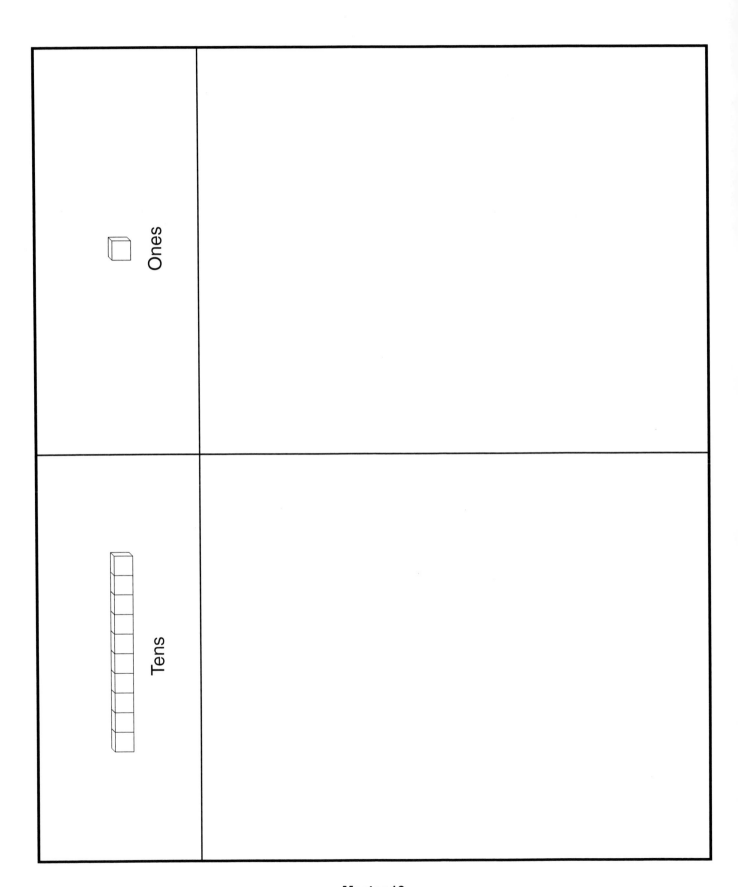

Master 10
© Math Teachers Press, Inc.
Reproduction only for one teacher for one class.

Place Value Mat (Part 2)

Hundreds

Thousands

Master 11
© Math Teachers Press, Inc.
Reproduction only for one teacher for one class.

Geoboard Dot Paper

1.

2.

3.

4.

5.

6.

7.

8.

9.

Master 12
© Math Teachers Press, Inc.
Reproduction only for one teacher for one class.

My Math Glossary

Write a definition or draw a picture for each word.

acute angle _____

addition _____

area _____

associative (grouping) property _____

average _____

bar graph _____

capacity _____

center _____

circle _____

circumference _____

closed figure _____

common factor _____

commutative (order) property _____

compass _____

congruent _____

coordinate points _____

data _____

decimal fraction _____

degrees (°) _____

denominator _____

diagonal _____

diameter _____

differences _____

discount _____

dividend _____

division _____

divisor _____

dozen _____

edge _____

equivalence _____

equivalent fractions _____

estimate _____

Master 13a
© Math Teachers Press, Inc.
Reproduction only for one teacher for one class.

My Math Glossary

Write a definition or draw a picture for each word.

face _____

factor _____

flip _____

fraction _____

greatest common factor _____

horizontal _____

improper fraction _____

intersecting _____

leading digit _____

least common denominator _____

line graph _____

line segment _____

lowest terms _____

mean _____

mixed number _____

multiplication _____

numerator _____

obtuse angle _____

open figure _____

ordered pair _____

parallel _____

parallelogram _____

percent _____

perimeter _____

perpendicular _____

pictograph _____

place value names to hundred billions place ____

place value decimal names to ten-thousandths place ___

price per unit _____

prime number _____

probability _____

Master 13b

© Math Teachers Press, Inc.
Reproduction only for one teacher for one class.

My Math Glossary

Write a definition or draw a picture for each word.

product _____

proper fraction _____

property _____

protractor _____

quadrilateral _____

quotient _____

radius _____

range _____

reciprocal _____

regular decagon _____

regular hexagon _____

regular octagon _____

regular pentagon _____

regular polygon _____

right angle _____

rounding _____

set _____

similarities _____

simplify _____

slide _____

square _____

straight angle _____

subtraction _____

sum _____

triangle _____

turn _____

unlike fractions _____

Venn diagram _____

vertex _____

vertical _____

volume _____

weight _____

Mathematics Table

LENGTH

Metric

1 kilometer = 1000 meters
1 meter = 100 centimeters
1 centimeter = 10 millimeters

Customary

1 mile = 1760 yards
1 mile = 5280 feet
1 yard = 3 feet
1 foot = 12 inches

CAPACITY & VOLUME

Metric

1 liter = 1000 milliliters

Customary

1 gallon = 4 quarts
1 gallon = 128 ounces
1 quart = 2 pints
1 pint = 2 cups
1 cup = 8 ounces

MASS AND WEIGHT

Metric

1 kilogram = 1000 grams
1 gram = 1000 milligrams

Customary

1 ton = 2000 pounds
1 pound = 16 ounces

TIME

1 millenium = 1000 years
1 decade = 10 years
1 year = 12 months
1 week = 7 days
1 hour = 60 minutes

1 century = 100 years
1 year = 365 days
1 year = 52 weeks
1 day = 24 hours
1 minute = 60 seconds

Vocabulary Cards

Master 15
© Math Teachers Press, Inc.
Reproduction only for one teacher for one class.

Name _____

The Billions Family

The family of three places after the millions family is the billions family.

[Place value chart showing Billions | Millions | Thousands | Units, each divided into Hundred, Ten, One; digits: 6 5 4, 2 8 1, 7 6 3, 0 2 9]

The billions family contains digits 6, 5 and 4.

The 4 is in the one billions place. It has a value of 4,000,000,000.
The 5 is in the ten billions place. It has a value of 50,000,000,000.
The 6 is in the hundred billions place. It has a value of 600,000,000,000.

Use the numeral above to answer questions 1–4.

1. What digits are in the millions family? _____ _____ _____
2. What digits are in the thousands family? _____ _____ _____
3. What digits are in the units family? _____ _____ _____
4. What digits are in the billions family? _____ _____ _____

Write the place name and value of the underlined digit.

5. <u>8</u>7,243,581,264 _____ _____

6. 9,54<u>8</u>,762,021 _____ _____

7. <u>1</u>,217,965,432 _____ _____

8. <u>9</u>35,462,000,000 _____ _____

For the number 216,354,809,207 write the digit that is in the place of the

9. one billions _____   10. ten thousands _____   11. hundred billions _____

12. hundred thousands _____   13. ten billions _____   14. tens _____

15. ones _____   16. ten millions _____   17. hundred millions _____

18. one millions _____   19. one thousands _____   20. hundreds _____

Skill Builders 1-1

Name _____

Use our mouths to compare numbers.
3510 > 3498

Use > or < to show how the numbers compare.

1. 7 ☐ 10
2. 84,365 ☐ 9,110
3. 901 ☐ 1,000

4. 843,024 ☐ 922,436
5. 1,230 ☐ 123
6. 99 ☐ 101

7. 8,432,127 ☐ 7,299,999
8. 27,263 ☐ 8,391

9. 106,000 ☐ 215,000
10. 9,873,000 ☐ 10,000,000

11. 28,295 ☐ 28,160
12. 931,000 ☐ 899,999

Order the numbers from least to greatest.

13. 13,872   13,849   13,881
_____   _____   _____

14. 1,254,986   1,254,843   1,254,879
_____   _____   _____

15. 19,125   20,006   19,365
_____   _____   _____

16. 870,643   870,622   870,980
_____   _____   _____

17. 2,643,596   2,727,128   2,643,569
_____   _____   _____

18. 55,783,296   55,783,297   55,783,299
_____   _____   _____

Skill Builders 2-1

Name _____

Rounding to the Nearest Thousand

You can use base ten blocks and a "halfway" number to help round numbers. Here is how it works...

Round 1843 to the nearest thousand.	
1. Build the number 1843.	**2.** Build the two groups of thousands between which 1843 falls. 1843 comes between 1000 and 2000.
3. Compare 1843 to 1000 and 2000 by thinking of a "halfway" number. 1500 is halfway between 1000 and 2000. 1843 is more than the halfway number. So 1843 rounds to 2000.	

Round each number to the nearest 1000.

1. 7322 comes between _____ and _____.

   The halfway number is _____.

   7322 is (greater, less) than 7500.

   So 7322 rounds to _____.

2. 4565 comes between _____ and _____.

   The halfway number is _____.

   4565 is (greater, less) than 4500.

   So 4565 rounds to _____.

	Number	Comes Between	Halfway #	Rounds To
3.	1058	<u>1000</u> and <u>2000</u>	<u>1500</u>	1058 rounds to <u>1000</u>
4.	9160			
5.	3440			
6.	9856			
7.	782			
8.	4742			
9.	333			
10.	12,110			
11.	39,850			
12.	115,200			

Skill Builders 3-1

Name _____

How did the clever dairy worker get a raise?

H E   B U T T E R E D   U p
3 8   10 6 9 9 8 2 8 4   6 1

T H E   B O S S .
9 3 8   10 7 5 5

Round these numbers to the nearest thousand. Circle the correct choice and put the letter of the correct choice above the blank that matches the number of the problem. The first one is done for you.

1. The highest recorded airspeed of a Concorde jet was 1,023 mph.
   1,023 rounds to — **(1,000) P** / 2,000 T

2. The highest point in the world, Mt. Everest in Nepal, Tibet, is 29,028 ft.
   29,028 rounds to — 29,000 R / 30,000 S

3. The average depth of the Pacific Ocean is 12,925 feet.
   12,925 rounds to — 12,000 A / 13,000 H

4. In 2000 there were 96,293,634 individual tax returns filed.
   96,293,634 rounds to — 96,293,000 R / 96,294,000 D

5. Minnesota collected 4,319 dollars in taxes for 1 month.
   4,319 rounds to — 3,000 T / 4,000 S

6. There are 3,711,519,319 dollar bills in circulation.
   3,711,519,319 rounds to — 3,711,519,000 U / 3,711,520,000 E

7. The shortest ocean distance from New York to Marseille, France, is 3,896 nautical miles.
   3,896 rounds to — 3,000 A / 4,000 O

8. In 1984 7,773,332 passenger cars were made in the United States.
   7,773,332 rounds to — 7,772,000 I / 7,773,000 E

9. The road mileage from Minneapolis to Los Angeles is 1,889 miles.
   1,889 rounds to — 1,000 S / 2,000 T

10. The highest mountain in the 48 continental states, Mt. Whitney in California, is 14,494 ft.
    14,494 rounds to — 14,000 B / 15,000 C

    (From the World Almanac 1986.)

Skill Builders 3-2
© Math Teachers Press, Inc.
Reproduction only for one teacher for one class.

Name _____

Prime Numbers: Only One Rectangular Shape

Cut out the squares at the bottom of the page.
Make rectangular shapes to decide whether each number is prime. If you can only make one rectangle, the number is prime.

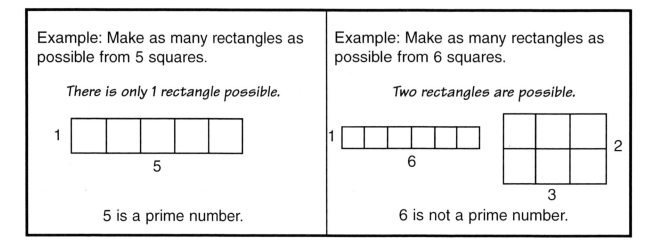

Example: Make as many rectangles as possible from 5 squares.

There is only 1 rectangle possible.

5 is a prime number.

Example: Make as many rectangles as possible from 6 squares.

Two rectangles are possible.

6 is not a prime number.

Investigate all the numbers in the chart and color those for which there is only 1 way to make a rectangle.

1	2	3	4	5	6	7	8	9	10
11	12	13	14	15	16	17	18	19	20
21	22	23	24	25	26	27	28	29	30
31	32	33	34	35	36	37	38	39	40

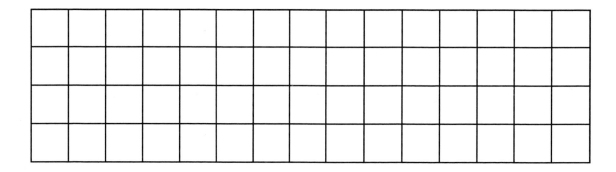

Skill Builders 4-1

Name _____

The Order and Associative Properties of Multiplication

The order (commutative) property	The grouping (associative) property
When we change the order of two numbers being multiplied, the product is the same. 5 × 3 = 15 3 × 5 = 15	When we multiply three numbers by grouping them in different ways, the product is the same. (2 × 3) × 5 = 2 × (3 × 5) 6 × 5 = 2 × 15 30 = 30 *Multiply the numbers in parentheses first.*

Use the order and associative properties to find the missing numbers.

1. If 7 × 8 = 56, 8 × 7 = _____

2. If 98 × 1094 = 107,212,
   1094 × 98 = _____

3. (8 × 6) × 2 = 8 × (_____ × 2)

4. 7 × _____ = 6 × 7

5. 800 × 8 = _____ × 800

6. (3 × _____) × 2 = 3 × (10 × 2)

7. (7 × 6) × 4 = 7 × (6 × 4)
   _____ × 4 = 7 × _____
   _____ = _____

8. 946 × 830 = 830 × _____

9. 15 × _____ = 5 × 15

10. _____ × (10 × 4) = (6 × 10) × 4

11. _____ × 67 = 67 × 31

12. 8 × _____ = 12 × 8

13. (2 × 9) × 6 = 2 × (9 × 6)
    _____ × 6 = 2 × _____
    _____ = _____

14. (_____ × _____) × 9 = 5 × (_____ × _____)
    20 × 9 = 5 × 36
    _____ = _____

Which property gives you a shortcut to multiply this problem?
(99,999 × 9,999) × 0 = _____

Skill Builders 5-1

Name _____

The Grouping or Associative Property

How do the answers compare when three numbers are multiplied in different groups?

Arnie has 4 shelves of shoes.
Each shelf has 3 pairs of shoes.
How many individual shoes does Arnie have?

There are 4 shelves.
Each shelf has (3 × 2) shoes.

4 × (3 × 2)
4 × 6 = ?

or,
3 pairs of shoes are on each of 4 shelves.

(3 × 2) × 4
6 × 4 = ?

Grouping or **associative property**: When multiplying three numbers, the result is the same no matter how the numbers are grouped.

4 × (3 × 2)	=	(4 × 3) × 2
4 × 6	=	12 × 2
24	=	24

1. (3 × 2) × 4 = 3 × (2 × 4)

   ___ × 4 = 3 × ___

   ☐ = ☐

2. 5 × (2 × 10) = (5 × 2) × 10

   5 × ___ = ___ × 10

   ☐ = ☐

3. (9 × 4) × 5 = 9 × (☐ × 5)

4. 7 × (3 × 4) = (☐ × 3) × 4

5. 3 × (10 × 2) = (3 × ☐) × 2

6. (7 × ☐) × 3 = 7 × (2 × 3)

7. (4 + 2) + 7 = 4 + (☐ + 7)

8. 9 + (6 + 1) = (9 + 6) + ☐

9. (6 + ☐) + 3 = 6 + (4 + 3)

10. ☐ + (4 + 5) = (6 + 4) + 5

11. Give an example of the grouping property of multiplication: _____

12. Give an example of the associative property of multiplication: _____

13. Which operations are associative?
    a. addition   b. subtraction   c. multiplication   d. division

14. Each classroom has 5 rows of desks with 6 desks in each row. There are 7 classrooms on the first floor. Write a number sentence to find the number of desks on the first floor. no. of desks = _____

Name _____

Add. The correct answers will show the path through the maze.

6 + 33,368 + 4,279 = _____

7,304 + 495 + 7 = _____

8,543 + 472 + 37 = _____

27 + 285,430 + 6,106 = _____

7 + 142 + 70,441 + 20 = _____

421 + 6,625 + 87,311 = _____

321,467 + 1,483 + 24 = _____

13,405 + 6 + 2,719 = _____

63 + 5,240 + 61,308 = _____

40,000 + 30,000 + 6 = _____

8 + 132 + 56,312 = _____

60,653 + 954 + 17 = _____

How do I get to my car?

	37,563	7907
37,653	7800	9152
7806	9052	291,561
78,601	70,601	291,563
71,609	70,610	93,457
73,222	74,357	94,357
16,135	16,231	322,974
66,666	16,130	66,315
66,611	70,116	66,713
56,542	70,006	56,651
61,623	56,452	61,533
61,624		

There are lots of paths... but the correct path adds up to 1,104,838.

Skill Builders 6-1
© Math Teachers Press, Inc.
Reproduction only for one teacher for one class.

Name _____

Subtracting 6-Digit Numbers

$$
\begin{array}{r}
537,862 \\
- 264,051 \\
\hline
\_\_3,811
\end{array}
$$

$$
\begin{array}{r}
{\scriptstyle 4\ 13} \\
5\cancel{3}7,862 \\
- 264,051 \\
\hline
273,811
\end{array}
$$

Find the differences. Check by adding.

1. 946,325
   − 683,210   Check.

2. 767,873
   − 284,525   Check.

3. 837,634
   − 265,297   Check.

4. 845,936
   − 53,642   Check.

5. 753,196
   − 82,743   Check.

6. 460,219
   − 37,039   Check.

7. 563,407
   − 171,228   Check.

8. 342,600
   − 121,329   Check.

Skill Builders 7-1

Name _____

Juanita plans to save a quarter every day for a year. How much money will she save in all?

A penny saved is a penny earned.

She will save 25¢ every day for 365 days. This is a multiplication problem.

```
   365
×   25
  1825
  7300
  9125¢    or $91.25
```

Find the products.

1. 800 × 46

2. 634 × 32

3. 238 × 26

4. 402 × 23

5. 186 × 92

6. 196 × 67

7. 786 × 85

8. 405 × 49

9. 967 × 49

10. 538 × 74

11. 796 × 83

12. 835 × 25

13. 17 × 237 = _____

14. 24 × 716 = _____

15. 245 × 32 = _____

16. 13 × 603 = _____

Skill Builders 8-1
© Math Teachers Press, Inc.
Reproduction only for one teacher for one class.

Name _____

```
____   ____   ____    ____
2248   2800   70,564  7752

____   ____   ____    ____
1806    540   39,780  3752

____   ____   ____
1806    540    7752

____   ____   ____    ____   ____   ____   ____   ____
3035   1806   23,352  24,219 7752   3752   1806   3035
```

> What is the most important thing for a teacher to know?

Solve each problem.
Write the letter of the problem in the blank above the answer.

A 1326
   × 30
   ____

D 897 × 27 = ____

E   408
   × 19
   ____

H 9 × 60 = ____

M 281
   × 8
   ____

N 67 × 56 = ____

O 400 × 7 = ____

R 3068
   × 23
   ____

S 607 × 5 = ____

T   86
   × 21
   ____

U 1946 × 12 = ____

The three best things about school: (1) June (2) July (3) August

Skill Builders 8-2
© Math Teachers Press, Inc.
Reproduction only for one teacher for one class.

Name _____

Dividing by a 1-Digit Number

Jack pasted 1824 stamps in a book. Each row had spaces for 8 stamps. How many rows did he fill?

This problem takes apart groups of equal size. It is a division problem.

```
    228  rows of stamps
8)1824
  16
   22
   16
    64
    64
     0
```

Remember DMS↓

Find the quotients. Use the four steps for long division.

1. 4)8408   2. 7)5243   3. 8)7534   4. 3)2775

5. 6)4583   6. 4)3507   7. 2)1503   8. 9)1287

9. 3)69,876   10. 4)98,622   11. 2)15,439   12. 6)98,405

13. 8)72,976   14. 7)27,645   15. 9)20,304   16. 5)17,210

17. 3)134,621   18. 4)231,752   19. 5)321,765   20. 2)976,432

Skill Builders 9-1
© Math Teachers Press, Inc.
Reproduction only for one teacher for one class.

Name _____

Dividing by Multiples of 10

There are three ways to divide this problem: 10)240		

1. Subtract 10s from 240.
 	
240 230 220
−10 −10 −10 etc.
230 220 210
 	
There are 24 tens in 240.
	2. Build 240. Share the block into 10 groups.
	
	☐ ☐ ||||
	
	Each group will have ||∷ in it.
		3. Work the problem using DMS↓ steps.
		
		   24
		10)240
		   200
		    40
		    40
		     0

Find the quotient without using paper and pencil. Draw a picture of your answer using ■ for hundreds, | for tens and • for ones.

1. 20)320   2. 30)360   3. 10)170   4. 20)2640

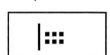 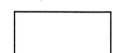 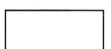 

Work the problems using the DMS↓ steps.

5. 40)560   6. 20)900   7. 50)650   8. 60)720

9. 80)960   10. 70)840   11. 30)450   12. 70)980

13. 30)7200   14. 40)1400   15. 40)1840   16. 50)7500

17. 60)2880   18. 50)1300   19. 80)1920   20. 90)3240

Skill Builders 10-1
© Math Teachers Press, Inc.
Reproduction only for one teacher for one class.

Name _____

Find the quotients. Shade in each region containing a correct answer.

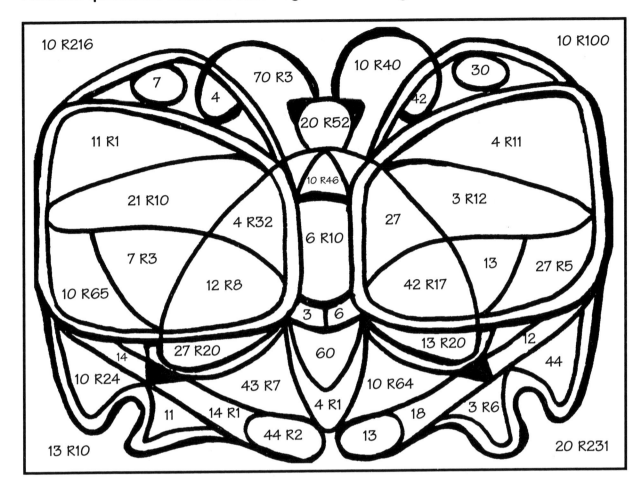

1. $52 \overline{)156}$
2. $41 \overline{)175}$
3. $71 \overline{)500}$
4. $21 \overline{)899}$

5. $32 \overline{)384}$
6. $12 \overline{)216}$
7. $46 \overline{)598}$
8. $63 \overline{)378}$

9. $23 \overline{)254}$
10. $33 \overline{)891}$
11. $67 \overline{)735}$
12. $32 \overline{)869}$

13. $73 \overline{)324}$
14. $12 \overline{)169}$
15. $11 \overline{)486}$
16. $73 \overline{)884}$

17. $37 \overline{)481}$
18. $42 \overline{)892}$
19. $35 \overline{)117}$
20. $54 \overline{)756}$

Skill Builders 10-2
© Math Teachers Press, Inc.
Reproduction only for one teacher for one class.

Name _____

Sometimes an estimated digit for the quotient is too large. Then try the next smaller digit.

Find the quotients.

1. 39)936
2. 28)938
3. 71)295
4. 25)725

5. 67)310
6. 29)619
7. 32)157
8. 45)270

9. 25)1287
10. 59)3068
11. 39)1683
12. 46)3818

13. 26)1924
14. 68)2636
15. 96)1097
16. 65)2665

17. – 32. Check problems 1–16 by multiplication.

Skill Builders 10-3
© Math Teachers Press, Inc.
Reproduction only for one teacher for one class.

Name _____

Naming Fractions

There are 12 inches in a foot. A bolt is 5 inches long. What fractional part of a foot is this bolt?

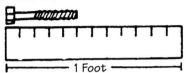

The bolt is $\frac{5}{12}$ of a foot.

There are 12 matching parts in this bar. There are 5 shaded parts. What fractional part of the bar is shaded?

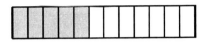

$\frac{5}{12}$ is shaded (five-twelfths).

1.

   How many matching parts? _____

   How many shaded parts? _____

   What fractional part
   is shaded? _____

2.

   How many matching parts? _____

   How many shaded parts? _____

   What fractional part
   is shaded? _____

Write the fraction for the shaded part of the bar.

3.
4.
5.
6.

7. ├──┼──┼──♦──┤
   0   X   1

   How many matching parts? _____

   How many parts from
   O to Point X? _____

   What fractional part of the line
   is represented by Point X? _____

8. ├─┼─┼─┼─♦─┼─┤
   0   X   1

   How many matching parts? _____

   How many parts from
   O to Point X? _____

   What fractional part of the line
   is represented by Point X? _____

What fractional part of the line does Point X represent?

9. ├─♦─┼──┼──┼──┤
   0 X   1

10. ├──┼──┼──♦──┤
    0   X   1

11. ├──┼──┼──┼──♦
    0   X

12. ├──┼──♦──┼──┤
    0   X   1

13. ├─┼─┼─┼─┼─♦─┼─┼─┤
    0   X   1

14. ├─┼─┼─┼─┼─┼─♦─┼─┼─┤
    0   X   1

Skill Builders 11-1

Name _____

Vocabulary of Fractions

The number above the fraction bar is called the **numerator**. The number below the fraction bar is called the **denominator**. $\frac{3}{4}$ numerator / denominator *Denominator and down both start with "D."*	A whole number with a fraction is called a **mixed number**. 
A **proper fraction** is less than a whole or 1. The numerator is less than the denominator. $\frac{3}{4}$ proper fraction *3 is less than 4.*	An **improper fraction** has a value of 1 or greater. $\frac{4}{4}$ and $\frac{5}{4}$ *Numerators are equal to or larger than denominators.*

1. Circle the fractions having a denominator of 4.

   $\frac{1}{4}$ $\frac{4}{5}$ 4 $\frac{5}{4}$

2. Circle the fractions having a numerator of 5.

   $\frac{4}{5}$ $\frac{5}{6}$ $\frac{5}{5}$ 5

3. Circle each fraction having a numerator greater than its denominator.

   $\frac{2}{3}$ $\frac{3}{3}$ $\frac{4}{2}$ $\frac{1}{8}$

4. Circle the proper fractions.

   $\frac{5}{5}$ $\frac{7}{12}$ $\frac{9}{8}$ $\frac{3}{10}$

5. Circle each fraction having a numerator equal to its denominator.

   $\frac{3}{3}$ $\frac{2}{2}$ $\frac{3}{2}$ 3

6. Circle the improper fractions.

   $\frac{5}{5}$ $\frac{4}{5}$ 2 $\frac{3}{11}$

7. Circle each fraction where the numerator is greater than the denominator.

   $\frac{5}{4}$ $\frac{10}{10}$ 3 $\frac{2}{3}$

8. Circle the improper fractions.

   $\frac{5}{6}$ 3 $\frac{4}{3}$ $\frac{10}{5}$

9. Circle the whole numbers.

   $\frac{2}{3}$ $\frac{7}{5}$ 2 $\frac{4}{6}$

10. Circle the mixed numbers.

   $\frac{8}{9}$ 2 $2\frac{1}{4}$ $\frac{12}{10}$

Skill Builders 11-2

Name _____

Equivalent Fractions in Lower Terms

The numerator and denominator of a fraction are called the **terms**. A fraction is changed to **lower terms** if the numerator of the new fraction is less than the original fraction.

$\frac{1}{3}$ and $\frac{2}{6}$ are equivalent fractions.

$\frac{1}{3}$ is in lowest terms.

$\frac{2}{6}$ is in higher terms.

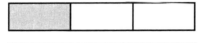

Rule: To change a fraction to lower terms, divide the terms (numerator and denominator) by the same number.

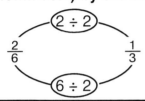

$\frac{1}{3}$ is in lower terms because the numerator 1 is less than the numerator 2.

Change to lower terms by dividing the numerator and denominator by 2.

1. $\frac{6}{8}$ _____ 2. $\frac{2}{4}$ _____ 3. $\frac{8}{12}$ _____ 4. $\frac{4}{10}$ _____

Change to lower terms by dividing the terms by 5.

5. $\frac{5}{10}$ _____ 6. $\frac{15}{20}$ _____ 7. $\frac{5}{5}$ _____ 8. $\frac{25}{30}$ _____

Change to lower terms by dividing the terms by 3 or 4.

9. $\frac{3}{9}$ _____ 10. $\frac{4}{12}$ _____ 11. $\frac{8}{12}$ _____ 12. $\frac{6}{9}$ _____

Change to lower terms by dividing the terms by a common number.

13. $\frac{6}{9}$ _____ 14. $\frac{2}{14}$ _____ 15. $\frac{12}{16}$ _____ 16. $\frac{8}{16}$ _____

17. $\frac{6}{15}$ _____ 18. $\frac{9}{24}$ _____ 19. $\frac{7}{21}$ _____ 20. $\frac{9}{27}$ _____

Skill Builders 12-1

Name _____

Simplest Fraction Bingo

1	$\frac{1}{6}$	$\frac{7}{8}$	$\frac{1}{3}$	$\frac{2}{5}$	$\frac{5}{9}$
$\frac{2}{7}$	$\frac{3}{4}$	$\frac{1}{9}$	$\frac{1}{2}$	1	$\frac{1}{7}$
$\frac{5}{8}$	$\frac{1}{2}$	1	$\frac{5}{6}$	$\frac{2}{9}$	$\frac{6}{7}$
$\frac{3}{5}$	$\frac{7}{9}$	$\frac{1}{4}$	$\frac{1}{3}$	$\frac{4}{5}$	$\frac{3}{7}$
$\frac{1}{4}$	$\frac{4}{7}$	$\frac{2}{3}$	1	$\frac{1}{8}$	$\frac{8}{9}$
$\frac{3}{8}$	1	$\frac{1}{5}$	$\frac{5}{7}$	$\frac{1}{2}$	$\frac{4}{9}$

A game for 2–5 players

Needed: Two 10-sided dice, a Fraction Bingo game board and markers of one color for each player.

Players take turns rolling the dice, forming a fraction and covering that fraction on the game board. The smaller number thrown becomes the numerator, and the larger number, the denominator; the fraction must be reduced to lowest terms. If 4 and 8 are thrown, the fraction $\frac{1}{2}$ would be formed. If 3 and 3 are thrown, the fraction reduces to 1. If a 0 is thrown, throw again. The first player to get six markers in a row in any direction—horizontal, vertical or diagonal—is the winner. To begin the play, each player throws the dice and forms a fraction; the player with the largest fraction begins.

Skill Builders 12-2

Name _____

Ordering Fractions

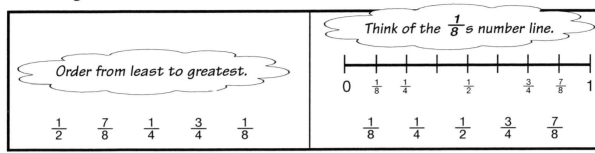

Order from least to greatest:

1. $\frac{1}{4}$ $\frac{4}{4}$ $\frac{3}{4}$ $\frac{0}{4}$ $\frac{2}{4}$

2. $\frac{3}{8}$ $\frac{1}{8}$ $\frac{7}{8}$ $\frac{4}{8}$ $\frac{0}{8}$

3. $\frac{5}{6}$ $\frac{1}{6}$ $\frac{1}{2}$ $\frac{2}{6}$ $\frac{6}{6}$

4. $\frac{10}{10}$ $\frac{1}{10}$ $\frac{1}{2}$ $\frac{7}{10}$ $\frac{3}{10}$

5. $\frac{4}{12}$ $\frac{0}{12}$ $\frac{7}{12}$ $\frac{12}{12}$ $\frac{1}{2}$

6. $\frac{1}{5}$ $\frac{5}{5}$ $\frac{3}{5}$ $\frac{1}{2}$ $\frac{2}{5}$

7. $\frac{1}{4}$ $\frac{1}{8}$ $\frac{1}{2}$ $\frac{3}{8}$ $\frac{5}{8}$

8. $\frac{1}{2}$ $\frac{3}{4}$ $\frac{1}{8}$ $\frac{7}{8}$ $\frac{5}{8}$

9. $\frac{1}{2}$ $\frac{1}{4}$ $\frac{1}{3}$ $\frac{1}{5}$ $\frac{1}{1}$

10. $\frac{1}{8}$ $\frac{1}{5}$ $\frac{1}{12}$ $\frac{1}{10}$ $\frac{1}{2}$

11. $\frac{1}{3}$ $\frac{1}{2}$ $\frac{1}{6}$ $\frac{2}{3}$ $\frac{0}{6}$

12. $\frac{2}{3}$ $\frac{1}{12}$ $\frac{1}{6}$ $\frac{1}{3}$ $\frac{1}{2}$

13. $\frac{1}{12}$ $\frac{5}{12}$ $\frac{1}{4}$ $\frac{1}{3}$ $\frac{1}{6}$

14. $\frac{1}{2}$ $\frac{3}{4}$ $\frac{7}{12}$ $\frac{2}{3}$ $\frac{11}{12}$

15. $\frac{2}{5}$ $\frac{3}{10}$ $\frac{1}{10}$ $\frac{3}{5}$ $\frac{1}{2}$

16. $\frac{7}{8}$ $\frac{1}{2}$ $\frac{1}{4}$ $\frac{3}{8}$ $\frac{3}{4}$

Skill Builders 13-1
© Math Teachers Press, Inc.
Reproduction only for one teacher for one class.

Mixed Numbers and Improper Fractions on a Number Line

Write each improper fraction or mixed number below its corresponding point on the correct number line. Write the letter from the code box above the point.

C	H	S	O	?	E	W	P	O	T	T	E	T	T	E	A	W	E	N
$\frac{7}{8}$	$\frac{10}{10}$	$\frac{4}{5}$	$\frac{15}{10}$	$\frac{13}{6}$	$\frac{15}{8}$	$\frac{2}{4}$	$1\frac{4}{8}$	$\frac{10}{6}$	$1\frac{5}{12}$	$\frac{12}{6}$	$\frac{7}{5}$	$1\frac{2}{8}$	$\frac{7}{6}$	$\frac{19}{10}$	$1\frac{3}{4}$	$\frac{7}{10}$	$\frac{8}{8}$	$\frac{9}{8}$

F	R	I	H	E	I	H	T	R	D	O	E	E	I	T	A	R	S	A
$\frac{11}{12}$	$\frac{9}{6}$	$\frac{8}{10}$	$\frac{5}{4}$	$\frac{15}{12}$	$\frac{11}{8}$	$1\frac{2}{6}$	$\frac{4}{4}$	$1\frac{7}{10}$	$1\frac{6}{8}$	$\frac{5}{5}$	$1\frac{1}{12}$	$\frac{13}{8}$	$\frac{3}{4}$	$\frac{9}{10}$	$1\frac{5}{6}$	$1\frac{1}{5}$	$\frac{13}{10}$	$\frac{5}{8}$

What's worse than a giraffe……

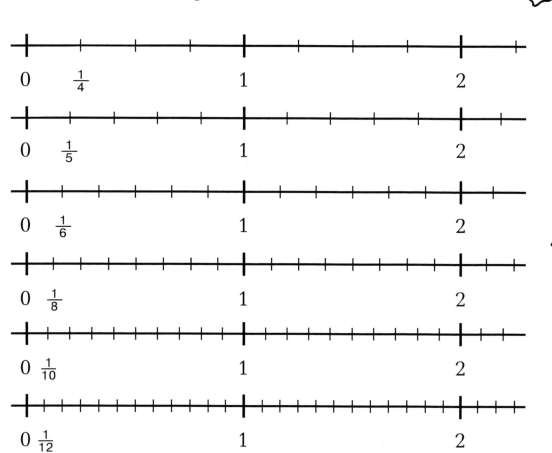

Skill Builders 14-1

Name _____

Subtracting Like Fractions and Simplifying to Lowest Terms

You can use fraction bars to subtract fractions with the same denominators.

Subtract the fraction pairs. Write the answer in lowest terms.

1. $\frac{5}{6} - \frac{1}{6}$

2. $\frac{7}{10} - \frac{3}{10}$

Subtract. Simplify to lowest terms.

3. $\frac{4}{6} - \frac{1}{6}$

4. $\frac{5}{10} - \frac{1}{10}$

5. $\frac{9}{12} - \frac{7}{12}$

6. $\frac{3}{6} - \frac{1}{6}$

7. $\frac{3}{12} - \frac{1}{12}$

8. $\frac{12}{15} - \frac{7}{15}$

9. $\frac{15}{30} - \frac{7}{30}$

10. $\frac{5}{20} - \frac{3}{20}$

11. $\frac{7}{9} - \frac{4}{9} =$ _____

12. $\frac{5}{8} - \frac{3}{8} =$ _____

13. $\frac{7}{8} - \frac{3}{8} =$ _____

14. $\frac{9}{16} - \frac{5}{16} =$ _____

15. Duane had $\frac{3}{4}$ of a pie. He ate $\frac{1}{4}$ of the pie. How much pie was left?

16. Peter's worm measures $\frac{9}{16}$ of an inch. His caterpillar is $\frac{13}{16}$ of an inch. How much longer is the caterpillar?

Skill Builders 15-1

Name _____

Adding Mixed Numbers and Simplifying

Add. Simplify to lowest terms.

1. $2\frac{1}{4}$
   $+ 1\frac{1}{4}$

2. $1\frac{3}{8}$
   $+ 2\frac{1}{8}$

3. $3\frac{1}{8}$
   $+ 1\frac{1}{8}$

4. $1\frac{1}{6}$
   $+ 1\frac{1}{6}$

5. $1\frac{3}{10}$
   $+ 1\frac{2}{10}$

6. $1\frac{2}{6}$
   $+ 2\frac{2}{6}$

7. $1\frac{7}{12}$
   $+ 2\frac{2}{12}$

8. $2\frac{5}{8}$
   $+ 1\frac{1}{8}$

9. $1\frac{3}{4}$
   $+ 1$

10. $3\frac{4}{9}$
    $+ \frac{2}{9}$

11. $2\frac{2}{15}$
    $+ 2\frac{8}{15}$

12. $\frac{3}{10}$
    $+ 4\frac{1}{10}$

13. $1\frac{1}{12}$
    $+ 2\frac{4}{12}$

14. $5\frac{1}{16}$
    $+ 2\frac{3}{16}$

15. $3\frac{6}{20}$
    $+ 2\frac{4}{20}$

16. $2\frac{5}{18}$
    $+ 1\frac{1}{18}$

17. Jan is making two batches of cookies. One calls for $2\frac{1}{4}$ cups of flour. The other calls for $3\frac{1}{4}$ cups flour. How much flour does Jan need in all?

18. On Monday the walking club walked 2 miles. On Wednesday they walked $2\frac{7}{10}$ miles. Friday they walked $2\frac{1}{10}$ miles. How far did they walk altogether?

Skill Builders 16-1
© Math Teachers Press, Inc.
Reproduction only for one teacher for one class.

Name _____

Subtracting Mixed Numbers with Exchanging

You can use fraction circles to help understand how to subtract mixed numbers.
You must remember that 1 whole circle has many fraction names: $\frac{2}{2}$, $\frac{3}{3}$, $\frac{4}{4}$, and so on.

| Build the larger number. Subtract the fraction part. Subtract the whole numbers.

$3\frac{1}{5}$
$-1\frac{3}{5}$

You cannot take $\frac{3}{5}$ from $\frac{1}{5}$. What can you do? | You must exchange 1 whole for $\frac{5}{5}$. Record the result.

$2\frac{6}{5}$
$-1\frac{3}{5}$
$\overline{1\frac{3}{5}}$ |

Use fraction circles. Record the exchange and the answer.

1. $3\frac{1}{8}$
   $-1\frac{6}{8}$

2. $4\frac{1}{4}$
   $-1\frac{2}{4}$

3. $3\frac{1}{3}$
   $-1\frac{2}{3}$

4. $3\frac{2}{5}$
   $-1\frac{4}{5}$

5. $3\frac{3}{10}$
   $-1\frac{6}{10}$

6. $4\frac{2}{6}$
   $-1\frac{3}{6}$

7. $3\frac{1}{12}$
   $-1\frac{6}{12}$

8. $3\frac{5}{8}$
   $-1\frac{2}{8}$

Find the differences.

9. $6\frac{2}{9}$
   $-1\frac{8}{9}$

10. $9\frac{7}{15}$
    $-3\frac{7}{15}$

11. $1\frac{4}{7}$
    $-\frac{5}{7}$

12. 12
    $-\frac{9}{10}$

13. $4\frac{1}{2}$
    -2

14. 4
    $-2\frac{5}{6}$

15. 6
    $-3\frac{1}{2}$

16. 5
    $-2\frac{1}{4}$

17. $3 - 1\frac{3}{8} =$ _____

18. $6 - 4\frac{2}{3} =$ _____

19. Kelly had 5 pizzas. Marty took $2\frac{1}{6}$ pizzas. How much was left for Kelly?

_____

20. A board measures $7\frac{3}{8}$ feet. A shelf $3\frac{7}{8}$ feet long is cut from the board. How long is the piece that is left?

_____

Skill Builders 16-2

Name _____

Adding Fractions with Different Denominators

To add fractions, they must name numbers with the same number of matching parts.

Add these fractions. Match the letters of each answer with the numbers to answer the question.

1. $\dfrac{1}{4} + \dfrac{2}{3}$ ☐ = G

2. $\dfrac{2}{3} + \dfrac{1}{6}$ ☐ = T

3. $\dfrac{2}{5} + \dfrac{1}{2}$ ☐ = L

4. $\dfrac{1}{4} + \dfrac{1}{3}$ ☐ = U

5. $\dfrac{3}{10} + \dfrac{2}{5}$ ☐ = I

6. $\dfrac{1}{6} + \dfrac{1}{3}$ ☐ = B

7. $\dfrac{2}{12} + \dfrac{1}{4}$ ☐ = T

8. $\dfrac{1}{2} + \dfrac{1}{6}$ ☐ = A

9. $\dfrac{2}{3} + \dfrac{3}{4}$ ☐ = O

10. $\dfrac{3}{4} + \dfrac{1}{3}$ ☐ = C

11. $\dfrac{2}{5} + \dfrac{7}{10}$ ☐ = H

12. $\dfrac{5}{8} + \dfrac{3}{4}$ ☐ = N

Where did King Arthur go for fun?

___ ___ ___ ___ ___ ___ ___ ___ ___ ___ ___ ___
$\dfrac{5}{6}$ $1\dfrac{5}{12}$ $\dfrac{2}{3}$ $1\dfrac{3}{8}$ $\dfrac{7}{10}$ $\dfrac{11}{12}$ $1\dfrac{1}{10}$ $\dfrac{5}{12}$ $1\dfrac{1}{12}$ $\dfrac{9}{10}$ $\dfrac{7}{12}$ $\dfrac{1}{2}$

Skill Builders 17-1
© Math Teachers Press, Inc.
Reproduction only for one teacher for one class.

Name _____

Adding and Subtracting Fractions with Different Denominators

Step 1. Find the L.C.D.
What is the smallest number both 6 and 4 will divide into evenly?

$6 \overline{)?}$ $4 \overline{)?}$

L.C.D. = 12

Step 2. Change fractions to equivalent fractions.

$\frac{1}{6} \underset{\times 2}{\overset{\times 2}{=}} \frac{2}{12}$

$\frac{3}{4} \underset{\times 3}{\overset{\times 3}{=}} \frac{9}{12}$

Step 3. Add.

$\frac{2}{12}$
$+\frac{9}{12}$
$\overline{\frac{11}{12}}$

Find the common denominator. Add.

1. $\frac{1}{6} + \frac{1}{4}$

2. $\frac{3}{5} + \frac{1}{3}$

3. $\frac{3}{8} + \frac{1}{6}$

4. $\frac{1}{9} + \frac{5}{6}$

5. $\frac{1}{5} + \frac{2}{7}$

6. $\frac{5}{8} + \frac{1}{3}$

7. $\frac{3}{5} + \frac{1}{6}$

8. $\frac{1}{12} + \frac{1}{8}$

Find the common denominator. Subtract.

9. $\frac{3}{4} - \frac{1}{6}$

10. $\frac{5}{6} - \frac{3}{8}$

11. $\frac{1}{2} - \frac{2}{11}$

12. $\frac{4}{5} - \frac{2}{3}$

13. $\frac{4}{9} - \frac{1}{3}$

14. $\frac{3}{4} - \frac{5}{9}$

15. $\frac{2}{3} - \frac{1}{12}$

16. $\frac{3}{5} - \frac{1}{4}$

Skill Builders 17-2
© Math Teachers Press, Inc.
Reproduction only for one teacher for one class.

Name _____

Adding Mixed Numbers with Unlike Denominators

To add mixed numbers, the fractions must be divided into the same number of matching parts. They must have the same denominators.

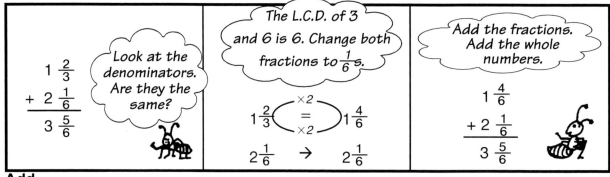

Add.

1. $2\frac{1}{2} + 1\frac{1}{4}$

2. $2\frac{1}{3} + 3\frac{1}{2}$

3. $1\frac{1}{8} + 2\frac{3}{4}$

4. $2\frac{3}{4} + \frac{1}{6}$

5. $2\frac{3}{8} + 2\frac{1}{2}$

6. $2\frac{1}{12} + 1\frac{4}{12}$

7. $2\frac{3}{10} + 1\frac{2}{5}$

8. $2\frac{1}{2} + 1\frac{2}{10}$

9. $4\frac{1}{5} + 2\frac{1}{3}$

10. $5\frac{1}{3} + \frac{3}{8}$

11. $3\frac{1}{6} + 2\frac{1}{8}$

12. $2\frac{1}{5} + 5\frac{2}{5}$

13. Juan walked $1\frac{7}{10}$ miles one day and $1\frac{1}{5}$ miles the next day. How far did he walk on both days? _____

14. One piece of material is $1\frac{1}{3}$ yards long. Another piece is $2\frac{1}{2}$ yards long. How many yards in all? _____

15. Ada caught two fish. One weighed $1\frac{3}{4}$ pounds and the other weighed $2\frac{1}{4}$ pounds. What was the total weight of the fish? _____

16. It rained $1\frac{5}{16}$ inches on Monday and $\frac{3}{8}$ inch on Tuesday. How much did it rain on both days? _____

Skill Builders 18-1
© Math Teachers Press, Inc.
Reproduction only for one teacher for one class.

Name _____

Multiplying Fractions

Cut each shaded part into two equal parts. Multiply.

1. $\frac{1}{2} \times \frac{1}{4} =$ _____

2. _____ × _____ = _____

3. _____ × _____ = _____

4. _____ × _____ = _____

Cut each shaded part into three equal parts. Multiply.

5. $\frac{1}{3} \times \frac{1}{4} =$ _____

6. _____ × _____ = _____

7. _____ × _____ = _____

8. _____ × _____ = _____

Cut each shaded part into four equal parts. Multiply.

9. $\frac{1}{4} \times \frac{1}{2} =$ _____

10. _____ × _____ = _____

Skill Builders 19-1
© Math Teachers Press, Inc.
Reproduction only for one teacher for one class.

Name _____

Multiplying Fractions and Whole Numbers

There are 12 eggs in a dozen.
Hal used $\frac{1}{3}$ of them in an omelet.
How many eggs did he use?

$\frac{1}{3}$ of 12 = ?

$\frac{1}{3}$ of 12 is the same as $\frac{1}{3}$ × 12.

$\frac{1}{3} \times 12 = \frac{1}{3} \times \frac{12}{1} = \frac{1 \times 12}{3 \times 1} = \frac{12}{3} = 4$

Find the missing numbers.

1. $\frac{1}{2} \times 6 = 6 \times$ _____

2. $\frac{3}{4} \times 8 =$ _____ $\times \frac{3}{4}$

3. $\frac{1}{2} \times$ _____ $= 10 \times \frac{1}{2}$

4. _____ $\times 9 = 9 \times \frac{2}{3}$

Find the products. Write the answers in lowest terms.

5. $\frac{1}{2} \times 8 =$ _____

6. $\frac{1}{3} \times 9 =$ _____

7. $\frac{1}{4} \times 12 =$ _____

8. $\frac{1}{5} \times 15 =$ _____

9. $\frac{1}{6} \times 18 =$ _____

10. $\frac{1}{8} \times 40 =$ _____

11. $\frac{1}{3} \times 24 =$ _____

12. $\frac{1}{7} \times 14 =$ _____

13. $\frac{3}{4} \times 16 =$ _____

14. $\frac{1}{2} \times 14 =$ _____

15. $\frac{3}{4} \times 100 =$ _____

16. $\frac{1}{4} \times 24 =$ _____

Choose the best estimate.

17. $\frac{1}{3} \times 4 =$ _____
(1, 2, 4, 12)

18. $\frac{1}{4} \times 9 =$ _____
(1, 2, 9, 36)

19. $\frac{1}{5} \times 19 =$ _____
(1, 4, 5, 19)

20. $\frac{1}{6} \times 32 =$ _____
(1, 5, 6, 32)

21. A recipe called for $\frac{1}{2}$ cup of butter. To double the recipe, how much butter would you use? _____

22. Doug had 80¢. He spent $\frac{1}{4}$ of it. How much did he have left? _____

Skill Builders 19-2
© Math Teachers Press, Inc.
Reproduction only for one teacher for one class.

Name _____

Reciprocals in Dividing Fractions

When you turn a fraction upside down, you are inverting the fraction.

$\frac{2}{3}$ inverted is $\frac{3}{2}$

Another name for an inverted fraction is **reciprocal**.

The reciprocal of $\frac{3}{4}$ is $\frac{4}{3}$.

To divide by a fraction, multiply by its reciprocal.

$20 \div \frac{1}{4} =$

$\frac{20}{1} \times \frac{4}{1} = \frac{80}{1} = 80$

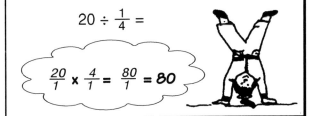

Invert each number.

1. $\frac{3}{4}$ _____
2. $\frac{4}{5}$ _____
3. $\frac{5}{1}$ _____
4. 7 _____

Write the reciprocal of each number.

5. $\frac{1}{3}$ _____
6. $\frac{5}{8}$ _____
7. $\frac{8}{1}$ _____
8. 4 _____

Divide.

9. $2 \div \frac{1}{3}$
10. $3 \div \frac{1}{4}$
11. $\frac{5}{8} \div 2$
12. $\frac{1}{8} \div 3$

13. $\frac{2}{3} \div \frac{3}{4}$
14. $\frac{1}{5} \div \frac{1}{3}$
15. $\frac{3}{4} \div \frac{4}{5}$
16. $\frac{1}{2} \div \frac{2}{3}$

17. $\frac{1}{7} \div \frac{2}{5}$
18. $\frac{3}{5} \div \frac{2}{3}$
19. $\frac{2}{5} \div \frac{9}{16}$
20. $\frac{4}{5} \div 3$

21. Tom baked 3 pies. He cut each pie into $\frac{1}{6}$ of a pie serving. How many servings were there? _____

22. A can of mixed nuts is 15 ounces. How many paper cups holding $\frac{3}{4}$ ounce each can be filled from the can? _____

Skill Builders 20-1
© Math Teachers Press, Inc.
Reproduction only for one teacher for one class.

Name _____

Decimal Fractions: Hundredths from Models

There are 100 pennies in 1 dollar. One penny is $\frac{1}{100}$ of a dollar. The value of a penny can be written: 1¢ or $\frac{1}{100}$ or $0.01	The large square is a whole or unit. It has been divided into 100 matching small squares. Each small square can be written: $\frac{1}{100}$ or 0.01 "one-hundredth" 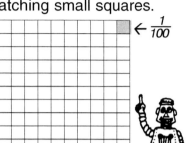

Write the value of the coins as a fractional part of a dollar and as a decimal.

1. _____ _____

2. _____ _____

Write a proper fraction and a decimal fraction for the shaded part of each figure.

3. _____ _____   4. _____ _____   5. _____ _____

Shade fractions equivalent to the numeral or words.

6. 0.03   7. 0.61   8. nineteen-hundredths

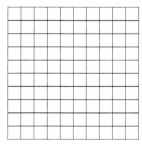 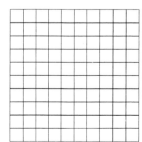 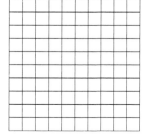

Skill Builders 21-1

Name _____

Writing Mixed Decimal Fractions

Read mixed decimal fractions the same way you read mixed proper fractions. Remember to say the word "and" between the whole number and fraction part.

1 whole unit

$\frac{3}{10}$ of a unit

1 and $\frac{3}{10}$ is shaded

1.3 is read "one <u>and</u> three-tenths"

Write a mixed fraction, a mixed decimal fraction and the words for the shaded figures.

1.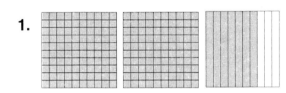

   _____ _____

   _____ _____

2.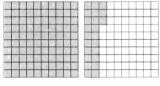

   _____ _____

   _____ _____

Write as mixed decimal fractions.

3. $1\frac{4}{10}$ _____   4. $5\frac{21}{100}$ _____   5. $2\frac{9}{10}$ _____   6. $3\frac{75}{100}$ _____

Write the words for the numerals. Remember, decimal words end in <u>ths</u>.

7. 0.13 _____

8. 0.5 _____

9. 1.7 _____

10. 6.12 _____

11. 21.7 _____

12. 4.05 _____

13. 0.30 _____

14. 9.4 _____

15. 2.61 _____

Skill Builders 22-1

Name ____

Decimal Place Value in Hundredths

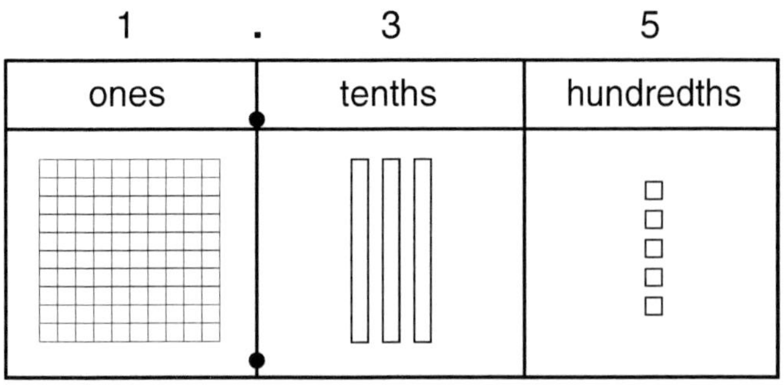

The second place to the right of the decimal point is called the hundredths place. The 5 is in the hundredths place.

Remember:
1 penny is 1¢
or $\frac{1}{100}$ of a dollar
or $0.01.

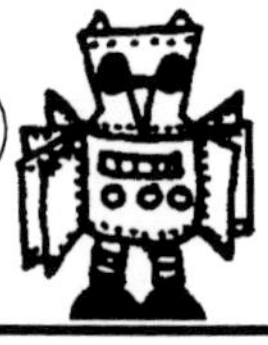

Circle the digit in the hundredths place.

1.

H	T	O	tenths	hundredths
		5	7	6

2.

H	T	O	tenths	hundredths
6	0	4	3	2

3. 0.71 **4.** 12.65 **5.** 3.04 **6.** 0.09

7. 60.53 **8.** 9.20 **9.** 14.651 **10.** 7.062

Write the place name for each underlined digit.

11. 6.<u>5</u>2 ____________________ **12.** 0.7<u>3</u> ____________________

13. 2<u>1</u>.64 ____________________ **14.** <u>3</u>5.71 ____________________

15. 253.7<u>1</u> ____________________ **16.** <u>7</u>06.59 ____________________

Write the numeral for each model.
Circle the digit in the tenths place in the numeral you have written.

17. ________

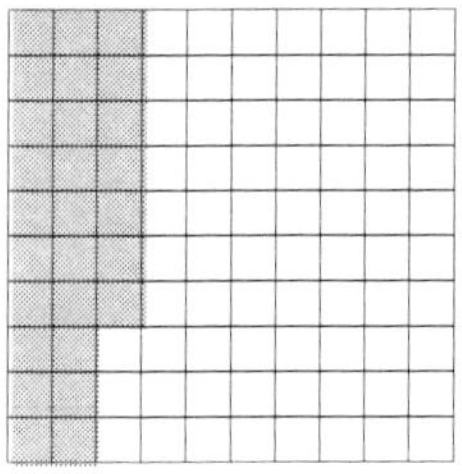

18. ________

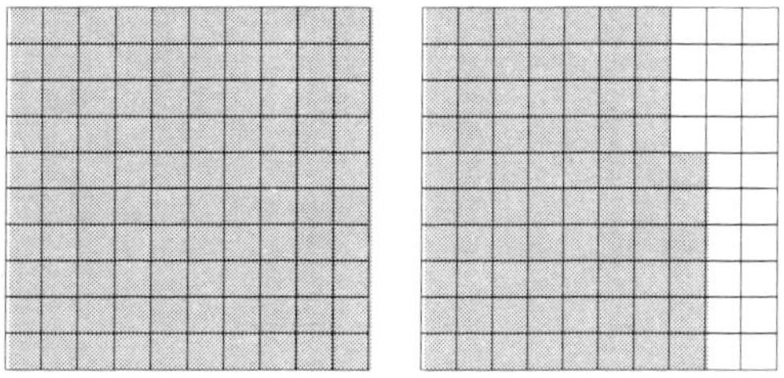

© Math Teachers Press, Inc.
Reproduction only for one teacher for one class.

Name _____

Decimal Place Value in Thousandths and Ten-Thousandths

The number 1.2357 is made up of 1 one, 2 tenths, 3 hundredths, 5 thousandths and 7 ten-thousandths.

$1.2357 = 1 + 0.2 + 0.03 + 0.005 + 0.0007$

ones	tenths	hundredths	thousandths	ten thousandths
1	2	3	5	7

Fill in the place value chart. The first one has been done for you.

DECIMAL POINT ↓

	Ten Thousands	Thousands	Hundreds	Tens	Ones	Tenths	Hundredths	Thousandths	Ten Thousandths
1. 6.531					6	5	3	1	
2. 27.089				2	7	0	8	9	
3. 31.6527				3	1	6	5	2	7
4. 372.9845			3	7	2	9	8	4	5
5. 507.906			5	0	7	9	0	6	
6. 0.9236					0	9	2	3	6

Write the place name for each underlined digit.

7. 29.34<u>6</u>   _____

8. 6.5<u>9</u>41   _____

9. <u>3</u>06.217   _____

10. 12.<u>2</u>96   _____

11. 6.354<u>8</u>   _____

12. <u>3</u>71.24   _____

13. <u>8</u>406.275   _____

14. 7.00<u>8</u>1   _____

15. 23.965<u>4</u>   _____

16. 0.6<u>5</u>32   _____

Skill Builders 23-2

Name _____

Comparing Decimal Fractions of Uneven Lengths

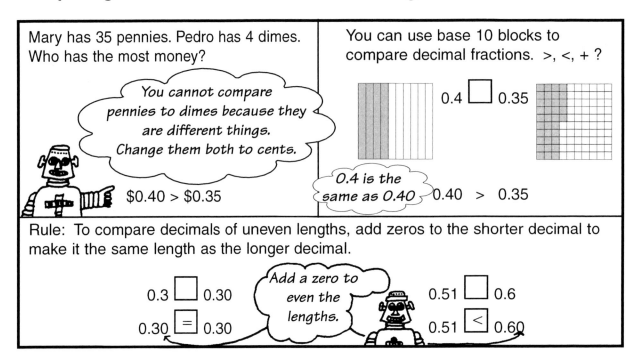

Compare the shaded amounts. Use >, <, or = to write a true statement.

1. 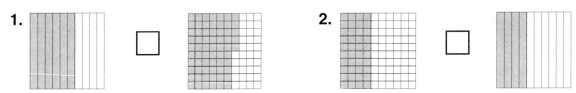   2.

Write >, <, = in the box to make a true statement?

3. 0.7 ☐ 0.68
4. 0.50 ☐ 0.5
5. 0.51 ☐ 0.6

6. 0.9 ☐ 0.90
7. 0.61 ☐ 0.5
8. 0.27 ☐ 0.5

9. 0.3 ☐ 0.38
10. 0.713 ☐ 0.684
11. 0.4 ☐ 0.004

12. 0.35 ☐ 0.352
13. 0.098 ☐ 0.9
14. 0.7 ☐ 0.08

Order from least to greatest.

15.   0.5   0.45   0.52   16.   0.09   0.4   0.2

_____ _____ _____   _____ _____ _____

Skill Builders 24-1
© Math Teachers Press, Inc.
Reproduction only for one teacher for one class.

Name _____

Write 1.3 as a fraction.	The last digit is in the tenths place. The denominator of the fraction must be 10.	$= 1\frac{3}{10}$
Write 0.67 as a fraction.	The last digit is in the hundredths place. The denominator of the fraction must be 100.	$= \frac{67}{100}$

Change each decimal to a fraction. Write the letter of the problem in the blank above the matching fraction.

(F) 0.21 = ____   (O) 0.69 = ____   (H) 0.37 = ____   (A) 0.83 = ____

(N) 1.7 = ____   (E) 1.1 = ____   (F) 1.03 = ____   (T) 0.3 = ____

(E) 1.27 = ____   (A) 0.7 = ____   (T) 0.9 = ____   (C) 0.70 = ____

(W) 1.0 = ____   (O) 0.03 = ____   (H) 0.09 = ____   (C) 1.07 = ____

When does 11 + 2 = 1?

___ ___   ___ ___ ___   ___ ___ ___ ___
$\frac{69}{100}$ $1\frac{7}{10}$ $\frac{9}{10}$ $\frac{37}{100}$ $1\frac{1}{10}$ $\frac{21}{100}$ $\frac{7}{10}$ $\frac{70}{100}$ $1\frac{27}{100}$

___ ___   A   ___ ___ ___ ___ ___   !
$\frac{3}{100}$ $1\frac{3}{100}$ $\frac{10}{10}$ $\frac{83}{100}$ $\frac{3}{10}$ $1\frac{7}{100}$ $\frac{9}{100}$

Skill Builders 25-1

Name _____

Adding and Subtracting Uneven Decimals

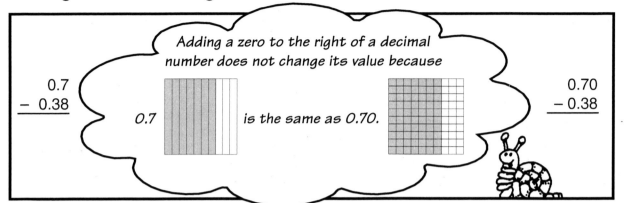

Work each problem. Find the answer in the code box. Put the letter of the answer in the blank above the problem number.

1. 0.86
   − 0.37

2. 8.4
   − 1.17

3. 6.9
   − 3.07

4. 7.25
   − 3.8

5. 57.03
   + 7.6

6. 14.7
   + 8.64

7. 58.9
   + 10.25

8. 3.17
   + 6.078

9. 8.00
   − 1.18

10. 5
   + 7.6

11. 6
   − 2.8

12. 0.25
   + 0.9

13. $4 − $1.40 = _____

14. 2.08 − 0.3 = _____

How do you make a 7 even?

___ ___ ___ ___   ___ ___ ___ ___
 5   14   2   10   1   8   6   4

   ___ ___ ___   ___ ___ ___
   13   7   11   12   3   9

Code Box

T	A	T	E	K	H	"	A	W	"	Y	E	S	A
64.63	23.34	2.60	3.2	7.23	69.15	6.82	0.49	9.248	1.15	3.45	12.6	3.83	1.78

Skill Builders 26-1

Name _____

Multiplying Decimals by a Whole Number

To multiply 4 × 0.2, think of putting together 4 groups of 2 tenths each.

0.2
× 4

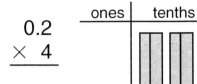

Say: 4 times 2 tenths is 8 tenths.

Write: 4 × 0.2 = 0.8

or

0.2
× 4
―――
0.8

1. 0.3
    × 2

2. 0.4
    × 3

3.   5
    × 0.3

4.   6
    × 0.3

5. 0.24
    × 2

6. 0.12
    × 6

7. 0.36
    × 7

8. 1.27
    × 8

9. $0.35
    ×   3

10. $1.95
    ×   8

11. $2.40
    × 12

12. $3.65
    × 14

13. Jack bought 8 apples for 24¢ each. How much did all the apples cost?

14. Tom bought 18 gallons of gasoline at 79¢ a gallon. How much did the gas cost?

15. Felicia bought 6 rose bushes for $14.75 each. How much did she pay for the roses?

16. A record costs $5.95. If you buy 5 of them for your friends, what is the total cost?

17. Baseball tickets cost $4.50 each. Find the cost of 4 tickets.

18. One shirt costs $14.00. Slacks cost $24.00. Find the total cost of 2 shirts and 2 pairs of slacks.

Name _____

Multiplying Decimals by Decimals

Here are two ways to multiply a decimal by a decimal:

change to fractions:	with a rule:
0.3 × 0.4 = ? $\frac{3}{10} \times \frac{4}{10} = \frac{12}{100} = 0.12$	Multiply. The number of decimal places (digits to the right of the decimal point) in the product is equal to the sum of the decimal places in the factors. 0.3 (1 place) 0.4 (1 place) 0.12 2 places

Change each decimal to a fraction and multiply. Write the answer as a fraction and as a decimal.

		Fractions	Fraction Answer	Decimal Answer
1.	0.3 × 0.2			
2.	0.5 × 0.3			
3.	0.7 × .31			
4.	0.32 × 0.04			

5.   6.2
  × 0.3

6.   0.24
  × 0.8

7.   0.24
  × 0.08

8.   0.15
  ×   4

9.   3.7
  × 0.05

10.   0.75
  × 0.4

11.   7.4
  × 0.9

12.   0.68
  × 0.07

13.   2.5
  × 0.12

14.   0.56
  × 0.24

15.   3.2
  × 1.5

16.   0.29
  ×   14

Skill Builders 27-2
© Math Teachers Press, Inc.
Reproduction only for one teacher for one class.

Name _____

Dividing Decimals by Whole Numbers

Lunch tickets cost $2.00 each. Five tickets can be purchased at a special price of 5 for $7.50. How much does one lunch ticket cost at the special price?

Use the DMS↓ steps to find the special price. Remember the $ sign and the decimal point in your answer.

```
      $1.50
   5 )$7.50
      5
      2 5
      2 5
         0
```

Find the quotients.

1. $2\overline{)0.54}$ 2. $3\overline{)0.87}$ 3. $5\overline{)0.65}$ 4. $6\overline{)0.72}$

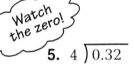

5. $4\overline{)0.32}$ 6. $8\overline{)0.48}$ 7. $9\overline{)0.45}$ 8. $5\overline{)7.20}$

9. $4\overline{)\$9.40}$ 10. $6\overline{)\$8.10}$ 11. $5\overline{)\$2.05}$ 12. $8\overline{)\$8.96}$

13. A package of six candy bars is marked $2.70. What is the cost of one candy bar?

    _____

14. Cass earned $13.00 for working 4 hours. What is his hourly wage?

    _____

15. A bunch of four bananas is marked 48¢. What is the cost of one banana?

    _____

16. A six-pack of soda pop is marked $2.40. What is the cost of one can of soda?

    _____

Skill Builders 28-1

Name _____

Dividing Decimals by Decimals

Craig sold homemade taffy for $0.06 a piece. He sold $7.50 of taffy after school one day. How many pieces of taffy did he sell?

```
    125
6 ) 750
    6
    ‾‾
    15
    12
    ‾‾
    30
    30
    ‾‾
     0
```

You can change both numbers to pennies and divide.

$0.06 = 6 pennies
$7.50 = 750 pennies

Rule: To divide by a decimal number, move the decimal point to the right to make it a whole number. Move the decimal point in the number being divided the same number of places. Add zeros if necessary. Divide. Place the decimal point in the quotient.

```
         125              125
0.06 ) 7.50          6 ) 750
```

2 places

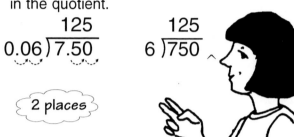

Find the quotients.

1. 0.06) 0.84   2. 0.03) 0.96   3. 0.08) 0.96   4. 0.05) 0.95

5. 0.9) 8.1   6. 0.4) 0.92   7. 0.2) 0.86   8. 0.7) 9.8

9. 0.07) 0.21   10. 0.9) 0.45   11. 4) 5.40   12. 0.2) 6.70

13. Pencils cost $0.08 each. How many pencils can be bought for $2.56?

14. Bubble gum costs $0.04 a piece. How many pieces can be bought for $2.00?

15. A piece of fabric 2.4 meters long is to be cut and made into t-shirts. Each t-shirt uses 0.3 meter of material. How many t-shirts can be made?

16. A large box of candy weighs 25.6 kilograms. The candy is to be weighed and put into sacks of 0.4 kilogram each. How many sacks will be used?

Skill Builders 28-2
© Math Teachers Press, Inc.
Reproduction only for one teacher for one class.

Name _____

The Meaning of Percent

Percent is a special way of comparing parts per hundred or hundredths. Here are two ways to think about percent.

1 penny has a value of 1 cent.
1 dollar has a value of 100 cents.
1 cent is $\frac{1}{100}$ of a dollar or 1% of a dollar.
Remember, "cent" is related to "hundredths," "percent" means "parts per hundred."

The large square has 100 small squares.
1 small square is shaded.

1 large square = $\frac{100}{100}$ = 100%
1 small square = $\frac{1}{100}$ = 1%

Fill in the blanks.

1. 3 pennies = ____¢ = ____%
2. 1 nickel, 2 pennies = ____¢ = ____%
3. 1 dime, 3 pennies = ____¢ = ____%
4. 1 quarter, 4 pennies = ____¢ = ____%

What percent do the following shaded amounts represent?

5. ____%
6. ____%
7. ____%

Shade in the small squares to represent the given percent.

8. 21%
9. 75%
10. 100%

Do these circles seem reasonable? Explain why or why not.

11.
12.
13.

_____   _____   _____

_____   _____   _____

Skill Builders 29-1

Name _____

Changing Percents and Fractions

A percent is another way to write fractional parts per hundred.

17% compares 17 parts per 100. 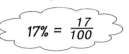	$\frac{23}{100}$ means 23 parts out of 100.

Write as a fraction.

1. 27% _____
2. 91% _____ (lowest terms)
3. 13% _____
4. 7% _____

5. 1% _____
6. 64% _____
7. 100% _____
8. 20% _____

9. 50% _____
10. 25% _____
11. 75% _____
12. 90% _____

Write as a percent.

13. $\frac{24}{100}$ _____
14. $\frac{50}{100}$ _____
15. $\frac{8}{100}$ _____
16. $\frac{10}{100}$ _____

17. $\frac{85}{100}$ _____
18. $\frac{1}{100}$ _____
19. $\frac{17}{100}$ _____
20. $\frac{69}{100}$ _____

21. $\frac{100}{100}$ _____
22. $\frac{1}{2}$ _____
23. $\frac{1}{4}$ _____
24. $\frac{3}{4}$ _____

Guess what percent of each circle is labeled A, B and C.

25. 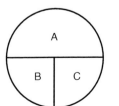 A = ____% B = ____% C = ____%

26. 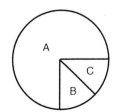 A = ____% B = ____% C = ____%

Skill Builders 30-1
© Math Teachers Press, Inc.
Reproduction only for one teacher for one class.

Name _____

Angle

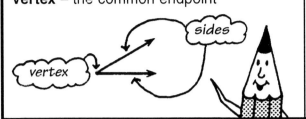
angle – two rays from a common endpoint
vertex – the common endpoint

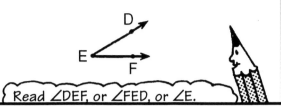
To name an angle, write the letter of the vertex in the middle.
Read ∠DEF, or ∠FED, or ∠E.

Circle the part of each object that contains an angle. If there are none, cross out the object.

1.
2.
3.
4.
5.

Name each angle in three ways.

6.

_____ _____ _____

7.

_____ _____ _____

8.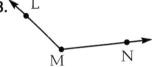

_____ _____ _____

Draw and label each figure.

9. ∠ABC

10. angle X

11. angle MNO

There are 6 rays drawn from the same point.

How many angles are there?

Count in a systematic way.

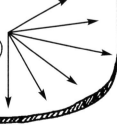

Skill Builders 31-1

Name _____

Perpendicular Lines

1. Fold a piece of paper like this. Trace the folded line with a pencil and straight edge.	2. Fold your paper like this. Trace the folded line. 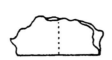	3. The two lines you traced form **perpendicular** lines.

1. Find two examples of perpendicular lines in your classroom.

    _____

2. Find two examples of parallel lines in your classroom.

    _____

3. Draw a line perpendicular to line m.

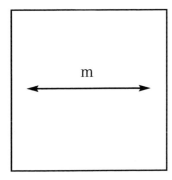

4. Draw a line perpendicular to line CD.

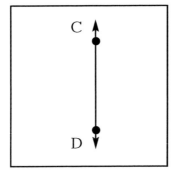

5. Draw a line perpendicular to line XY.

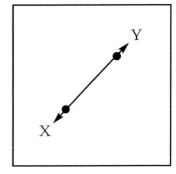

6. $\overline{CD}$ is _____ to $\overline{BD}$

   $\overline{CD}$ is _____ to $\overline{AB}$

7. $\overline{MO}$ is _____ to $\overline{NP}$

   $\overline{MO}$ is _____ to $\overline{MN}$

8. Can you name 4 pairs of perpendicular lines in the figure?

   _____ _____
   _____ _____

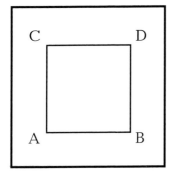

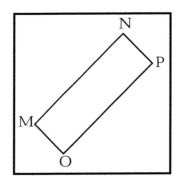

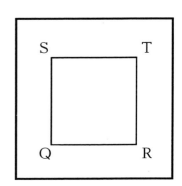

Name _____

Right Angle, Acute Angle, Obtuse Angle

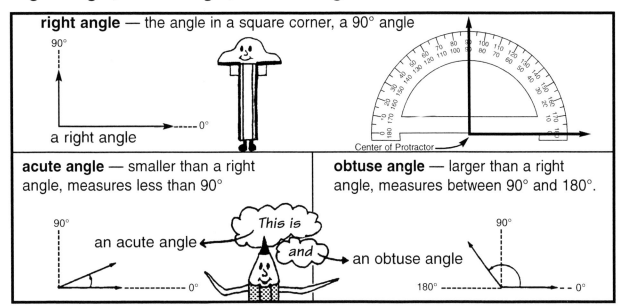

What kind of angle is suggested: acute, right or obtuse?

1. the open blades of scissors _____

2. corners of the American flag _____

3. corners of a stop sign _____

4. hands of clock at 3:00 _____

What kind of angle? Write A for acute, R for right and O for obtuse.

5.

6.

7.

8.

9.

10.

11. The clock hands at 5:00 _____

12. The clock hands at 3:20 _____

13. The clock hands at 6:15 _____

Skill Builders 33-1
© Math Teachers Press, Inc.
Reproduction only for one teacher for one class.

Name _____

Naming Polygons

A **polygon** is a closed, straight-line figure. These are polygons.

These are not polygons. Can you tell why?

A **triangle** is a polygon with three sides.
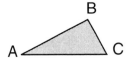
This figure is named triangle ABC or △ABC.

A **quadrilateral** is a polygon with four sides.
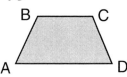
This figure is called ABCD.

A **pentagon** has five sides.

This is a regular pentagon.

A **hexagon** has six sides.

This is a regular hexagon.

An **octagon** has eight sides.

This is a regular octagon.

A **decagon** has ten sides.

This is a regular decagon.

Name the polygon.

1. _____
2. _____
3. _____
4. _____

5. _____
6.  _____
7. _____
8. _____

9. What quadrilateral is a regular polygon?

10. How many angles and vertices in a pentagon? _____

11. How many vertices and sides in an octagon? _____

12. What is the least number of line segments needed to form a polygon? _____

Skill Builders 34-1

Name _____

Parts of a Circle

A **compass** is an instrument used to draw circles. The point of the compass becomes the **center** of the circle. The curved line made by the pencil is the **circumference**.	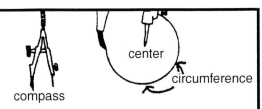 compass, center, circumference
A line segment from the center to any point on the circumference is called a **radius**. $\overline{XY}$ is a radius of the circle.	A line segment passing through the center with endpoints on the circumference is called a **diameter**. 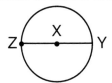 $\overline{ZY}$ is a diameter of the circle.

What part of each circle is represented by the dotted lines?

1. _____

2. 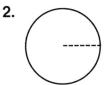 _____

3. _____

What part of the circle is suggested by:

4. the rim of a circular wastebasket?

   _____

5. the spokes of a wheel?

   _____

Give the radius and diameter of each circle.

6.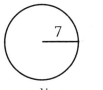

   radius = _____

   diameter = _____

7.

   radius = _____

   diameter = _____

8.

   radius = _____

   diameter = _____

9. Each chair on a Ferris wheel is 20 feet from the center. What is the greatest distance between any two chairs?

   _____

10. $\overline{CD}$ and $\overline{CE}$ are radii of the same circle. How do they compare in length?

    _____

Skill Builders 35-1
© Math Teachers Press, Inc.
Reproduction only for one teacher for one class.

Name _____

Making and Using a Ruler

1. Cut out the 1 inch measure. Use it to mark off 1-inch lengths along the edge of the ruler. Label each mark 0, 1, 2, 3, and so on.

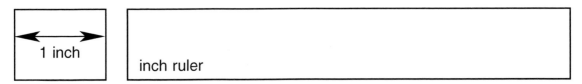

2. Cut out your ruler. Measure the lines below. Write the measure above each line.

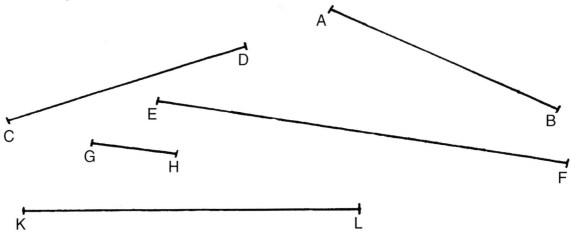

3. Cut out the $\frac{1}{2}$ inch measure. Use it to mark off $\frac{1}{2}$ inch lengths along the edge of the ruler. Label each mark 0, $\frac{1}{2}$, 1, and so on.

4. Cut out your ruler. Measure the lines below to the nearest $\frac{1}{2}$ inch. Write the measure above each line.

2 inches
2 in.
2"

all mean the same measure

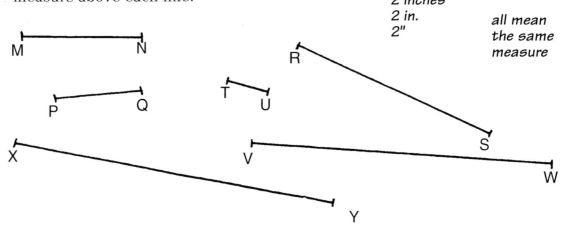

Skill Builders 36-1
© Math Teachers Press, Inc.
Reproduction only for one teacher for one class.

Name _____

Measuring in $\frac{1}{8}$s

Label each mark on the ruler with its simplest name (0, $\frac{1}{8}$, $\frac{1}{4}$, $\frac{3}{8}$, $\frac{1}{2}$, etc.). Cut out the ruler. Measure the lines to the nearest $\frac{1}{8}$ inch. Write the measure above each line.

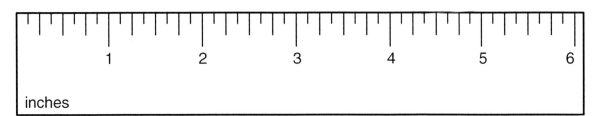

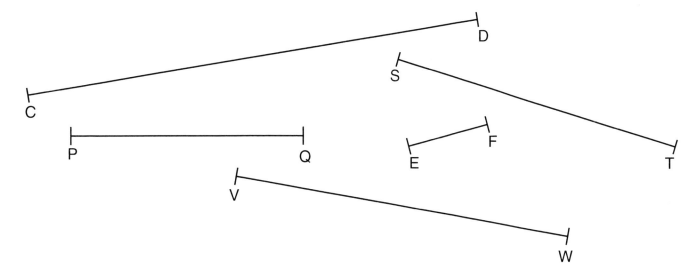

Count by $\frac{1}{8}$s from $\frac{0}{8}$ to $\frac{24}{8}$:

$\frac{0}{8}$, ___, ___, ___, ___, ___, ___, ___, ___,

___, ___, ___, ___, ___, ___, ___, ___,

___, ___, ___, ___, ___, ___, ___, $\frac{24}{8}$.

Count by $\frac{1}{8}$s from 0 to 2. Write the simplest name for each number.

0, $\frac{1}{8}$, $\frac{1}{4}$, ___, ___, ___, ___, ___, 1,

$1\frac{1}{8}$, ___, ___, ___, ___, $1\frac{3}{4}$, ___, 2.

Skill Builders 36-2
© Math Teachers Press, Inc.
Reproduction only for one teacher for one class.

Name _____

Drawing Angles with a Protractor

Use a protractor to draw and measure angles. The protractor has two scales, an inner and outer.

The angle is read on the inner scale because one ray of the angle rests on 0° on the inner scale. What is the measure of the angle?

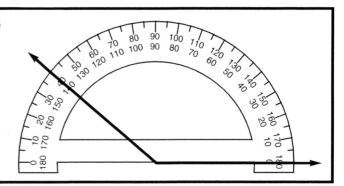

Use a protractor to draw angles using the rays and degrees given.

1.

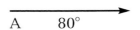
A   80°

2.

X   45°

3.

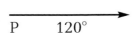
P   120°

4.

T   60°

5.

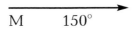
M   150°

6.

R   30°

Skill Builders 37-1
© Math Teachers Press, Inc.
Reproduction only for one teacher for one class.

Name _____

Use a ruler or compass to find the perimeter.

1.

P = _____ cm

2.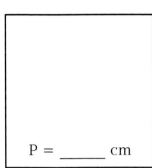

P = _____ cm

3.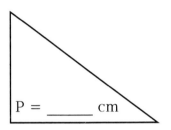

P = _____ cm

4.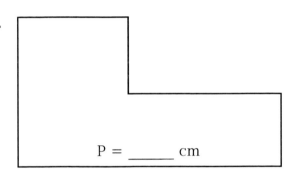

P = _____ cm

5.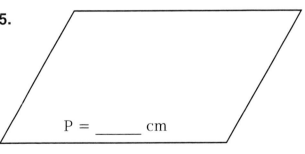

P = _____ cm

6.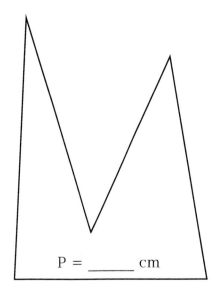

P = _____ cm

Skill Builders 38-1

Name _____

Finding Area by Counting Unit Squares

Each small square is 1 unit square. Give the area of each figure in square units.

1. Area = _____ sq. units   **2.** Area = _____ sq. units   **3.** Area = _____ sq. units

 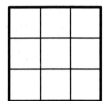 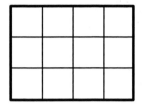

4. Area = _____ sq. units   **5.** Area = _____ sq. units

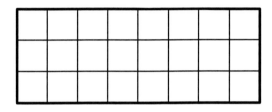 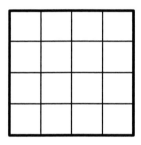

6. Area = _____ sq. units   **7.** Area = _____ sq. units   **8.** Area = _____ sq. units

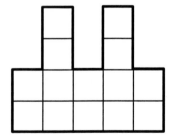 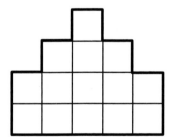 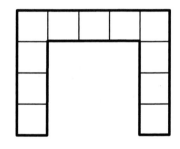

9. Area = _____ sq. units   **10.** Area = _____ sq. units   **11.** Area = _____ sq. units

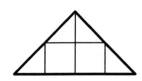 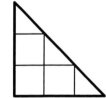 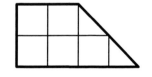

Skill Builders 38-2
© Math Teachers Press, Inc.
Reproduction only for one teacher for one class.

Name _____

Filling Boxes

1. Here is a pattern for a box that is 5 centimeters × 4 centimeters × 1 centimeter. Cut out the pattern and tape the sides.

   How many centimeter cubes does it take to fill the box?

   _____ cu. cm

2. Make a pattern for a box that is 4 centimeters × 3 centimeters × 2 centimeters. Cut and tape the sides.

   How many centimeter cubes does it take to fill the box?

   _____ cu. cm

Use centimeter graph paper.

3. Make a pattern for a box that is 10 centimeters × 10 centimeters × 1 centimeter. How many centimeter cubes does it take to fill the box?

4. Use two sheets of centimeter graph paper to make a pattern for a box that is 10 centimeters × 10 centimeters × 10 centimeters. How many centimeter cubes does it take to fill the box?

Skill Builders 39-1

Name _____

Faces, Edges and Vertices of Solid Figures

Here is a pattern that can be folded to make a cube.

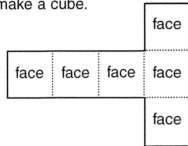

Each of the squares on the sides, top and bottom of the cube are called **faces**.

An **edge** is a line where any two faces meet.
A **vertex** is a point where any two edges meet.

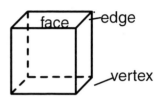

A cube has 6 faces, 12 edges and 8 vertices.

Here is a rectangular solid and a pattern for a rectangular solid.

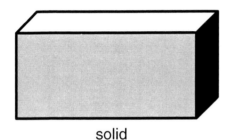
solid

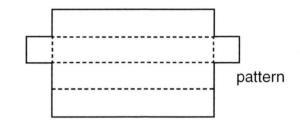
pattern

1. How many faces are there in a rectangular solid? _____

2. What shapes are the faces of a rectangular solid? _____

3. How many edges? _____

4. How many vertices? _____

Here is a pyramid and a pattern for a pyramid.

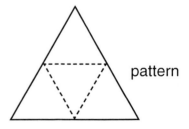
pattern

5. How many faces are there in a pyramid? _____

6. Describe the shapes of the faces of a pyramid. _____

7. How many edges? _____

8. How many vertices? _____

Skill Builders 39-2
© Math Teachers Press, Inc.
Reproduction only for one teacher for one class.

Name _____

Converting Time

1 year = 12 months	1 day = 24 hours
1 year = 52 weeks	1 hour = 60 minutes
1 week = 7 days	1 minute = 60 seconds

Use the table to answer the questions.

1. 2 hours = ____ minutes
2. 180 seconds = ____ minutes
3. 3 years = ____ months
4. 5 weeks = ____ days
5. 2 days = ____ hours
6. 130 minutes = ____ hours
7. 1 hr. 30 min. = ____ min.
8. 2 hr. 12 min. = ____ min.
9. 2 hr. 40 min. = ____ min.
10. 5 hr. 10 min. = ____ min.
11. 30 min. = ____ hr. (lowest terms)
12. 45 min. = ____ hr. (lowest terms)
13. 90 min. = ____ hr. (lowest terms)
14. 200 min. = ____ hr. (lowest terms)

Add. Simplify your answer.

15. 2 hr. 15 min.
   + 3 hr. 30 min.

16. 1 hr. 20 min.
   + 2 hr. 30 min.

17. 7 hr. 45 min.
   + 1 hr. 15 min.

Subtract.

18. 4 hr. 30 min.
   − 2 hr. 10 min.

19. 3 hr. 55 min.
   − 1 hr. 28 min.

20. 3 hr.
   − 1 hr. 15 min.

What is your favorite store?
How many hours is it open each day? in a week?

Skill Builders 40-1
© Math Teachers Press, Inc.
Reproduction only for one teacher for one class.

Units of Weight

The gram and kilogram are used to measure weight in the metric system.

One paper clip weighs about 1 gram

One textbook weighs about 1 kilogram

1000 grams (g) = 1 kilogram (kg)

A bowling ball weighs 7 kg. How many grams does the bowling ball weigh?

1 kg is 1000 g, so 7 kg is 7 x 1000 or 7000 g.

Would you measure in grams or kilograms?

1. a tennis ball _____
2. a television set _____
3. a thumbtack _____
4. a student in your class _____
5. a cat _____
6. a dime _____

Complete.

7. 2 kg = _____ g
8. 3000 g = _____ kg
9. 4 kg = _____ g
10. $3\frac{1}{2}$ kg = _____ g
11. 2500 g = _____ kg
12. 1300 g = _____ kg
13. 500 g = _____ kg
14. $1\frac{1}{4}$ kg = _____ g
15. 200 g = _____ kg
16. 2.3 kg = _____ g
17. 1750 g = _____ kg
18. 3.62 kg = _____ g

There are 1000 meters in a kilometer and 1000 grams in a kilogram.
What does the word "kilo" mean?
How many liters do you think there are in a kiloliter?

Skill Builders 41-1
© Math Teachers Press, Inc.
Reproduction only for one teacher for one class.

Name _____

Customary Units of Capacity

Tables of Capacity

1 pint (pt.) = 2 cups (c.)
1 quart (qt.) = 2 pints (pt.)
1 quart (qt.) = 4 cups (c.)
1 gallon (gal.) = 4 quarts (qt.)

3 gallons = _____ quarts

1 gallon has 4 quarts, 2 gallons has 8 quarts... Multiply.

Complete.

1. 2 pt. = ____ c.
2. 2 qt. = ____ c.
3. 2 pt. = ____ qt.

4. 1 gal. = ____ pt.
5. 4 pt. = ____ qt.
6. 2 gal. = ____ qt.

7. 16 qt. = ____ gal.
8. 3 pt. = ____ c.
9. 1 gal. = ____ c.

10. $\frac{1}{2}$ pt. = ____ c.
11. $1\frac{1}{2}$ qt. = ____ c.
12. $1\frac{1}{2}$ pt. = ____ c.

13. 6 c. = ____ qt.
14. $1\frac{1}{2}$ gal. = ____ qt.
15. 2 qt. = ____ gal.

16. A bathtub holds 32 quarts of water. How many gallons does it hold?

17. Janice made 6 quarts of punch. How many cups of punch is this?

18. Ernie bought 8 gallons of gas at $0.85 a gallon. How much did the gas cost?

19. Mr. Horn made a gallon of lemonade for a picnic. How many cups of lemonade did he make?

Name _____

Making Change

Jacinta bought a cake and a pie.
How much change should she get from a $5.00 bill?

Cookies $.72/bag
Pie $2.35
Cake $1.90

$ 1.90   1.
+ 2.35
―――――
  4.25

$ 5.00   2.
− 4.25
―――――
   .75

She should receive $.75 or 75¢ in change.

1. Kelly bought a bag of cookies. How much change did she get from a $1.00 bill?

2. Carlos bought a cake. How much change did he get from a $5.00 bill?

3. Stacy bought a pie. How much change did she get for a $10.00 bill?

4. Paula bought a bag of cookies and a cake. How much change did she get for a $10.00 bill?

5. Seung bought 2 bags of cookies. How much change should he get for a $5.00 bill?

6. Brian bought 2 cakes and a pie. How much change did he get for a $10.00 bill?

7. Carol bought 3 pies and 2 cakes. How much change did she get for a $20.00 bill?

8. Kim bought 2 bags of cookies and a pie. How much change did she get for a $10.00 bill?

9. Afifeh bought a pie and a bag of cookies. How much more did she pay for the pie?

10. Todd bought a bag of cookies and a cake. Noah bought a pie. How much more did Todd spend?

Skill Builders 43-1

Name _____

Patterns

Count from 156 to 192 by 3 and connect the dots in order.

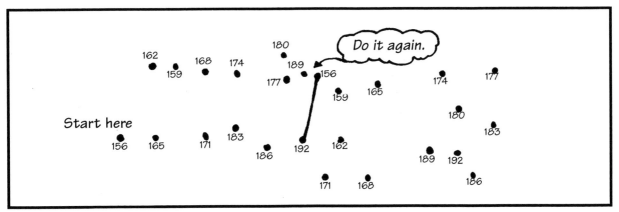

Fill in the missing numbers in the pattern.

1. 7 ▷ 11 ▷ 15 ▷ 19 ▷ ▷ ▷ ▷
2. 67 ▷ 62 ▷ 57 ▷ 52 ▷ ▷ ▷ ▷
3. 85 ▷ 78 ▷ 71 ▷ ▷ ▷ ▷ ▷
4. 0 ▷ 6 ▷ 12 ▷ 18 ▷ ▷ ▷ ▷
5. 2 ▷ 3 ▷ 5 ▷ 7 ▷ 11 ▷ ▷ ▷
6. 1 ▷ 1 ▷ 2 ▷ 3 ▷ 5 ▷ ▷ ▷

These patterns combine addition and subtraction.

4, 7, 5, 8, ____, ____, ____, ____, 8

9, 7, 12, 10, ____, ____, ____, ____, 21

Skill Builders 44-1
© Math Teachers Press, Inc.
Reproduction only for one teacher for one class.

Name _____

Deciding on a Process
In solving word problems, decide whether you should add, subtract, multiply or divide. Choose the correct number sentence to solve each problem and to complete the puzzle.

Daffy-nition of a drama teacher:

___ ___ ___ ___ ___   ___ ___ ___ ___ ___
 1   3   5   7   9   2   4   6   8   10

1. There are 18 students in a math class. When time was called, 11 had finished a timed test. How many were still working? _____
   - I. 18 + 11
   - S. 18 − 11
   - T. 18 x 11
   - H. 18 ÷ 11

2. Paper comes in packs of 500 sheets. How many sheets in 18 packs? _____
   - R. 500 + 18
   - W. 500 − 18
   - C. 500 x 18
   - D. 500 ÷ 18

3. Kirsten was helping the teacher. 186 pencils were needed. Each box held 12 pencils. How many boxes are needed? _____
   - H. 186 + 12
   - E. 186 − 12
   - A. 186 x 12
   - T. 186 ÷ 12

4. A cassette holds 240 minutes of film. How many 60-minute shows fit on the cassette? _____
   - E. 240 + 60
   - A. 240 − 60
   - L. 240 x 60
   - O. 240 ÷ 60

5. Justin counted 246 red notebooks and 467 gold notebooks. How many notebooks were there? _____
   - A. 246 + 467
   - E. 467 − 246
   - R. 467 x 246
   - O. 467 ÷ 246

6. Concert tickets cost $6. How many tickets can be bought with $54? _____
   - G. 6 + 54
   - R. 54 − 6
   - N. 54 x 6
   - A. 54 ÷ 6

7. Mt. McKinley, Alaska, is 20,320 ft. high. Eagle Mountain, Minnesota, is 2301 ft. How much taller is Mt. McKinley? _____
   - T. 20,320 + 2301
   - G. 20,320 − 2301
   - M. 20,320 x 2301
   - L. 20,320 ÷ 2301

8. Babysitting pays $3 an hour. How much will be earned for 6 hours? _____
   - M. 3 + 6
   - N. 6 − 3
   - C. 6 x 3
   - K. 6 ÷ 3

9. Sara bought a tape for $9.49 and a novel for $4.95. How much did she spend? _____
   - E. 9.49 + 4.95
   - L. 9.49 − 4.95
   - A. 9.49 x 4.95
   - S. 9.49 ÷ 4.95

10. In 1984 women over 55 watched an average of 42 hours of TV per week. Teenage girls watched an average of 21 hours per week. What is the difference in viewing time? _____
    - T. 42 + 21
    - H. 42 − 21
    - E. 42 x 21
    - K. 42 ÷ 21

Skill Builders 45-1
© Math Teachers Press, Inc.
Reproduction only for one teacher for one class.

Name _____

Steps in problem-solving process:	Some strategies for solving word problems:
1. Read and understand. 2. Circle the facts. Underline the question. 3. Decide on a process. 4. Estimate the answer. 5. Solve. Compare.	1. Act out the problem. 2. Simplify the numbers. 3. Draw a picture. 4. Use models to represent the information. 5. Make a chart to find a pattern.

Read the problem and identify the strategies (1–5 above) that could be used to solve the problem. Follow the steps to answer the questions.

1. A hot dog vendor sold 12 trays of hot dogs. There were 18 hot dogs on each tray. How many hot dogs were sold?

   Strategies? _____

   What is the process? _____

   Estimate _____ Answer _____

2. John's uncle gave him 22 stamps which makes the total in his collection 378. How many stamps did John have originally?

   Strategies? _____

   What is the process? _____

   Estimate _____ Answer _____

3. Sue took an 11-day trip to Alaska. She planned to spend $85 a day for gas. How much will she spend for gas?

   Strategies? _____

   What is the process? _____

   Estimate _____ Answer _____

4. Sue bought 272 postcards on her trip. She used 8 postcards on each page of a scrapbook. How many pages did she fill?

   Strategies? _____

   What is the process? _____

   Estimate _____ Answer _____

5. Four vendors sold 15 hot dogs, 28 hot dogs, 41 hot dogs and 39 hot dogs. How many hot dogs were sold in all?

   Strategies? _____

   What is the process? _____

   Estimate _____ Answer _____

6. A car averages 30 miles per gallon of gas. How many gallons will be needed to travel a distance of 480 miles?

   Strategies? _____

   What is the process? _____

   Estimate _____ Answer _____

Skill Builders 45-2

Name _____

Solving Word Problems with Fractions

1. There are 25 students in a class. $\frac{2}{5}$ of the class received an "A" on a test. How many students received an "A"?

2. There are 12 trails in the city park. Ernie hiked 8 of them. What fraction of the trails has Ernie hiked? (lowest terms)

3. Ernie met 15 other hikers on his walk. Nine of them wore headphone radios. What fraction of the walkers he met wore headphones? (lowest terms)

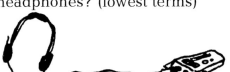

4. One hole is drilled $\frac{9}{16}$ inch in diameter. Another hole is $\frac{7}{16}$ inch in diameter. How much larger is the first hole? (lowest terms)

5. Hanna had $\frac{3}{4}$ of a granola bar. She ate $\frac{1}{2}$ of that. What part of the whole bar did she eat?

6. A pattern calls for $3\frac{1}{4}$ yards for a jacket, $1\frac{1}{4}$ yards for a shirt and 1 yard for a vest. How much material is needed to make all three?

7. Three people divided $\frac{3}{4}$ of a cake equally. What fractional part of the cake does each person get?

8. Sue cut a piece of material $3\frac{3}{4}$ yards long from a piece 10 yards long. How much was left?

Skill Builders 45-3

Name _____

Finding the Better Buy

To find the unit price of one item, divide the selling price by the number of items in a package.

Which package of batteries is the better buy?

 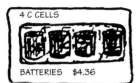

A battery in a 2 pack costs $1.19.
A battery in a 4 pack costs $1.09.
The 4-pack battery is the better buy.

Write your answers in complete sentences.

---

1. Which is the better buy: a box of 10 greeting cards for $7.50 or a box of 25 cards for $15.00? How much better?

2. Which is the better buy: a pack of 6 frozen juice bars for $1.80 or a single juice bar for 35¢? How much better?

3. Which is the better buy: a 10-ounce bag of potato chips for $2.09 or a 2-ounce bag for $0.48? How much better?

4. Which is the better buy: a single candy bar for 45¢ or a package of 6 candy bars for $2.10? How much better?

5. Which is the better buy: 50 napkins for $0.75 or 100 napkins for $1.00? How much better?

6. Which is the better buy: a 12-ounce jar of peanut butter for $1.20 or a 40-ounce jar of peanut butter for $3.60? How much better?

7. Which is the better buy: $\frac{1}{2}$ dozen doughnuts for $2.40 or 1 doughnut at $0.45? How much better?

8. Which is the better buy: 3 oranges for $1.05 or a dozen oranges for $3.60? How much better?

Skill Builders 45-4

Name _____

Problem Solving: Writing an Open Number Sentence, Simplifying

You can learn a general way to solve problems by writing number sentences. Use a box or letter to stand for the unknown number or **variable**.
A number sentence with a letter or variable is called an **open number sentence**.

1. Read.
2. Find question and the needed facts.
3. Decide on process.
4. Estimate.
5. Solve and check back.

Band members must sell 950 tickets to reach their goal. If 345 tickets have been sold, how many more tickets must be sold to reach the goal?

Choose a letter to stand for the unknown number or variable.

Putting in smaller numbers helps you decide on a process.
If 3 + n = 9, n = 6...
So you subtract to solve the problem.

Let n = number of tickets

Write a number sentence.

n = 950 − 345
  = 605 more tickets

Check: 345 + 605 = 950
    950 = 950

Write an open number sentence for each problem. Check.

1. Paul drives a truck. He uses one gallon of gas every 6 miles. If Paul has 13 gallons of gas in the tank and drives until he runs out of gas, how far did he drive?

   Let n = number of _____

   Number sentence _____

   n = _____

   Check: _____

2. Twelve people earned $584 collecting aluminum cans. What is the "fair share" each person should receive?

   Let n = number of _____

   Number sentence _____

   n = _____

   Check: _____

3. Tom has a collection of 26 piggy banks. David has a collection of 18 piggy banks. Find how many more piggy banks Tom has than David.

   Let n = number of _____

   Number sentence _____

   n = _____

   Check: _____

4. On the first day the Mall of America was open, 1,708,652 people visited the mall. The next day 1,700,600 people visited the mall. How many visitors were at the mall in these two days?

   Let n = number of _____

   Number sentence _____

   n = _____

   Check: _____

Skill Builders 45-5
© Math Teachers Press, Inc.
Reproduction only for one teacher for one class.

Name _____

Finding an Average Number with Manipulatives

The average of a group of numbers is the number that comes closest to representing the values of all the other numbers. The average number will be greater than some of the numbers and less than some others.

Jane has 5 pencils. The lengths of the pencils are 5 cm, 9 cm, 13 cm, 12 cm and 11 cm. What is the average length of the pencils?

- 5cm
- 9cm
- 13cm
- 12cm
- 11cm

Find the average length of the following sets of pencils. Build each pencil with 1 centimeter unit blocks or squares. Adjust the pencil lengths until they are all equal.

1. Pencils of 5 cm, 9 cm, 13 cm, 12 cm and 11 cm. Average length: _____ cm	2. Pencils of 9 cm, 5 cm, 15 cm and 11 cm. Average length: _____ cm
3. Pencils of 3 cm, 7 cm and 2 cm. Average length: _____ cm	4. Pencils of 8 cm, 6 cm, 10 cm, 12 cm and 4 cm. Average length: _____ cm
5. Pencils of 12 cm, 8 cm, 9 cm and 7 cm. Average length: _____ cm	6. Pencils of 18 cm, 12 cm, 14 cm, 13 cm and 23 cm. Average length: _____ cm

Skill Builders 46-1

Name _____

This chart shows Andy's scores on 5 tests.
To find Andy's average...

ANDY'S TEST SCORES
| 5 | 4 | 3 | 5 | 3 |

1. Find the total of the scores.

   5 + 4 + 3 + 5 + 3 = 20

2. Divide the total by the number of scores.

   $$5\overline{)20} = 4$$

My average test score is 4.

Solve each problem. Write the letter of the problem above the answer.

A. In 5 basketball games, Suzy scored 13, 8, 12, 10 and 17 points. Find her average. _____

B. Clark's test scores were 59, 78 and 49. What was his average score? _____

C. Find the average of these weights: 140, 142, 152, 138. _____

E. Find the average temperature for three days: 34°, 48°, 50°. _____

F. Eve's spelling scores were: 12, 17, 20, 19 and 12. Find the average. _____

G. Find the average temperature: 48°, 58°, 38°. _____

H. Scores on reading tests were: 88, 77, 92 and 75. Find the average. _____

L. Blake earned $8, $12.40, $7.60 and $16 by mowing lawns. What was his average pay? _____

M. Find the average weight: 136 lbs, 119 lbs, 108 lbs. _____

N. Marcy sold pop at the games. Her sales one month were 278 cans, 316 cans, 184 cans and 94 cans. Find her average sales. _____

O. John earned $23, $25 and $27 working part-time 3 days. Find his average earnings. _____

R. Kelly sold 252 newspapers in 3 days. What is her average sales per day? _____

S. One week the fifth grade class had the following absences: 8, 7, 4, 6 and 10. What was the average number of absences per day? _____

T. Les worked at the car wash. He recorded the cars washed each day. There were 96, 85, 97, and 82 cars. Find the average numbers of cars per day. _____

V. Find the average of these numbers: 18, 22, 19 and 13. _____

What can you say about a man with one hand on a hot stove and the other on a block of ice?

__ __ __ __ __   __ __ __ __ __ __ __ __
25 218 90 83 44   12 18 44 84 12 48 44

__ __ __   __ __ __ __ __ __ __ __ __ __ .
83 44 7   143 25 121 16 25 84 90 12 62 11 44

Skill Builders 46-2
© Math Teachers Press, Inc.
Reproduction only for one teacher for one class.

Name _____

Reading Charts

Calorie Chart

Egg, fried	100
Egg, poached	80
Bread, 1 slice	70
Butter, 1 T	100
Muffins, 1	156
Pancake	90
Syrup, 1 T	90
Orange juice	85

Use the data given in the table to solve these problems.

1. How many calories are there in 1 poached egg, 1 slice of bread and orange juice?

    _____

2. How many calories are there in 2 pancakes, 2 T of syrup, 2 T of butter and orange juice?

    _____

3. How many calories are there in a breakfast of 2 fried eggs, a glass of orange juice and a slice of toast with butter?

    _____

4. How many calories are there in 1 muffin, 1 T butter and 1 glass of orange juice?

    _____

5. Ingrid wants to eat over 400 calories because she wants to gain weight. What could she eat for breakfast?

    _____

    _____

6. Jake wants to eat less than 200 calories because he wants to lose weight. What could he eat?

    _____

    _____

7. A person breathes 24 times per minute. How many breaths does this make a day?

    _____

8. A bicycle burns 20 calories for each minute it is ridden. How long must you ride to burn the calories in 3 pancakes with 3 T of syrup?

    _____

Skill Builders 47-1
© Math Teachers Press, Inc.
Reproduction only for one teacher for one class.

Name _____

Probability: A Fair Game?

Predict
Play this game once. Decide whether the player is (equally likely, more than likely or less than likely) to win.

I think this game is (fair, unfair).

Game for 2 players

Each participant starts with 10 points.

The player rolls a pair of 6-sided dice ten times. If a sum of 7 is thrown, the opponent gives 3 points to the player. If any other sum is thrown, the player gives 1 point to the opponent. Record the number of points for the player and opponent at the end of each round.

The player with the most points at the end of the 10 throws is the winner. If a participant runs out of points, the other participant is the winner.

		1	2	3	4	5	6	7	8	9	10
Player	10										
Opponent	10										

Gather Data
Play this game many times. Gather data from other teams playing the game.

1. Describe the trend: The player is _____ going to win.

Explain
There are 36 ways to throw sums when using two dice.
Find the probability of throwing a sum of 7.

+	1	2	3	4	5	6
1						
2						
3						
4						
5						
6						

2. The chances of throwing a sum of 7 are _____ /36 or _____ /6

3. The chances of throwing a sum other than 7 are _____ /36 or _____ /6

4. Can you suggest a way to make this a fairer game?

_____

_____

Skill Builders 47-2

Name _____

Line Graphs

Jackie took a typing class for 10 weeks in the summer. She used a line graph to record her words per minute (wpm) on each weekly test.

1. How many words per minute did she type in Week 1?

   _____

2. How many words per minute (wpm) did she type in Week 5?

   _____

3. How many more wpm did she type in week 5 than in Week 1?

   _____

4. On which weeks did she drop in score?

   _____

5. Estimate the average improvement in wpm each week.

   _____

6. How many more wpm could she type in Week 10 than in Week 1?

   _____

Record the noon temperature outside for two weeks. Use the information to make a line graph.

Skill Builders 48-1
© Math Teachers Press, Inc.
Reproduction only for one teacher for one class.

Name _____

Estimating Sums

Approximate numbers are often used to estimate an answer. Rounded numbers are approximate numbers.

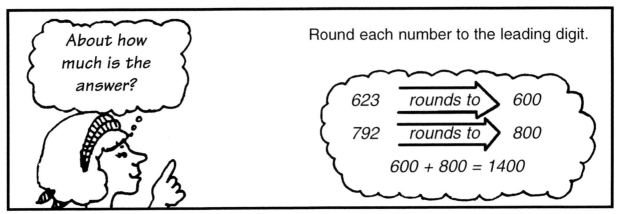

1. Which will give the best estimate for 81 + 68? ☐
    A. 80 + 60   **B.** 90 + 70   **C.** 70 + 70   **D.** 80 + 70

2. Which will give the best estimate for 427 + 684? ☐
    A. 400 + 600   **B.** 500 + 700   **C.** 300 + 700   **D.** 400 + 700

Use estimated sums to complete each table.

3.

+	21	59
38		
52		
47		

4.

+	32	66
17		
85		
74		

5.

+	370	637
210		
380		
450		

6.

+	420	590
364		
274		
158		

7.

+	1815	6184
4216		
3724		
2549		

8.

+	1827	4281
5178		
3547		
7862		

Skill Builders 49-1

Name _____

Estimating Differences

About how much is the answer?

7321
− 2970

Round each number to the leading digit in your head. Subtract the rounded numbers.

7321 rounds to 7000
2970 rounds to 3000
7000 − 3000 = 4000

1. Which will give the best estimate for 78 − 31? ☐
   A. 70 − 30   **B.** 80 − 40   **C.** 80 − 30   **D.** 80 − 20

2. Which will give the best estimate for 649 − 301? ☐
   A. 600 − 300   **B.** 700 − 400   **C.** 500 − 200   **D.** 700 − 300

3. Which will give the best estimate for 4616 − 3182? ☐
   A. 5000 − 4000   **B.** 4000 − 3000   **C.** 5000 − 2000   **D.** 5000 − 3000

Estimate the differences by rounding each number in the subtraction problem. Do not work the problem.

4. 81 → 80
   − 38 → − 40

5. 65 →
   − 21 → −

6. 89 →
   − 47 → −

7. 416 →
   − 154 → −

8. 728 →
   − 264 → −

9. 856 →
   − 371 → −

10. 6548 →
   − 2719 → −

11. 7184 →
   − 3826 → −

12. 8614 →
   − 3567 → −

Skill Builders 49-2
© Math Teachers Press, Inc.
Reproduction only for one teacher for one class.

Name _____

Estimating Products

Estimate the products by filling in the blanks.

1. To estimate 7 × 48, find the product of 7 × _____ = _____

2. To estimate 8 × 53, find the product of 8 × _____ = _____

3. To estimate 9 × 78, think of 9 × _____ = _____

4. To estimate 86 × 6, think of _____ × 6 = _____

5. To estimate 4 × 96, think of 4 × _____ = _____

6. To estimate 73 × 3, think of _____ × 3 = _____

7. To estimate 8 × 45, think of 8 × _____ = _____

8. To estimate 17 × 6, think of _____ × 6 = _____

9. To estimate 287 × 2, think of _____ × 2 = _____

10. To estimate 3 × 315, think of 3 × _____ = _____

11. 7 × 81: _____ × _____ = _____   12. 6 × 23: _____ × _____ = _____

13. 4 × 19: _____ × _____ = _____   14. 203 × 6: _____ × _____ = _____

15. 84 × 5: _____ × _____ = _____   16. 3 × 89: _____ × _____ = _____

17. 9 × 395: _____ × _____ = _____ 18. 7 × 812: _____ × _____ = _____

19. Vicki can pack 57 tapes in a box. Estimate the number of tapes she can pack in 6 boxes.
    Est. _____

20. A packing box holds 19 cups of styrofoam. Estimate the number of cups in 9 boxes.
    Est. _____

Skill Builders 50-1
© Math Teachers Press, Inc.
Reproduction only for one teacher for one class.

Name _____

Estimate the products by rounding the numbers to the leading digit and multiplying the rounded numbers.

	rounds to	estimate		rounds to	estimate
1. 21×49	____ × ____	_____	**2.** 48×9	____ × ____	_____
3. 28×53	____ × ____	_____	**4.** 34×83	____ × ____	_____
5. 63×75	____ × ____	_____	**6.** 56×32	____ × ____	_____
7. 64×47	____ × ____	_____	**8.** 9×83	____ × ____	_____
9. 68×75	____ × ____	_____	**10.** 55×42	____ × ____	_____
11. 71×300	____ × ____	_____	**12.** 49×400	____ × ____	_____
13. 91×286	____ × ____	_____	**14.** 64×512	____ × ____	_____
15. 38×406	____ × ____	_____	**16.** 25×634	____ × ____	_____
17. 17×289	____ × ____	_____	**18.** 85×397	____ × ____	_____
19. 46×165	____ × ____	_____	**20.** 67×570	____ × ____	_____

Complete the table by estimating the products.

21.

×	13	38
29		
41		
65		

22.

×	64	300
25		
78		
62		

23.

×	725	816
18		
43		
35		

Skill Builders 50-2
© Math Teachers Press, Inc.
Reproduction only for one teacher for one class.